Tomart's Encyclopedia of

Action Figures

Tomart's
Encyclopedia of
Action Figures

The 1001 Most Popular Collectibles of All Time

by Sally Ann Berk, Tom Tumbusch
and the editors of
TOMART'S ACTION FIGURE DIGEST

BLACK DOG
& LEVENTHAL
PUBLISHERS
NEW YORK

PUBLISHED BY

Black Dog & Leventhal Publishers, Inc.
151 West 19th Street
New York, NY 10011

DISTRIBUTED BY

Workman Publishing Company
708 Broadway
New York, NY 10003

Design by Martin Lubin Graphic Design

Typesetting by Brad Walrod/ High Text Graphics, Inc.

Manufactured in Spain

ISBN: 1-57912-009-1

h g f e d c b a

Library of Congress Cataloging-in-Publication Data

Berk, Sally Ann.
Tomart's encyclopedia of action figures: the 1001 most popular collectibles of all time/by Sally Ann Berk, Tom Tumbusch and the editors of Tomart's action figure digest.
     p.   cm.
ISBN 1-57912-009-1
1. Action figures (Toys)—Collectors and collecting—United States—Catalogs. I. Tomart's action figure digest.   II. Title.
NK8475.M5B47   1999
688.7'28'075—dc21   98-23244
                CIP

INTRODUCTION

The Barbie doll was introduced in 1959. The initial look didn't take, but the concept of a fashion doll was a huge success. Mattel redesigned Barbie as "the dream girl next door" and has not yet stopped reeling in the cash for ever-expanding sales. Major successes in the toy business don't go unnoticed. Merril Hassenfeld of Hasbro put the challenge to his creative staff. Don Levine, then head of developing boys' toys for Hasbro, considered several alternatives. He settled on a new toy soldier concept that would be bigger, more poseable, and capable of wearing many different outfits. Levine planned to go to market with one soldier for each major branch of U.S. military service—Army, Navy, Marines, and Air Force. These new "action fighters" all eventually ended up with the same name—G.I. Joe.

The military category was also useful by design to dispel any notion that these new "action figures" were dolls. No self-respecting boy would play with dolls, but positioning G.I. Joe as an action figure proved not only acceptable, but an immediate sales success rivaling that of Barbie. While Barbie was just another type of doll, G.I. Joe created a whole new category of boys' toys, one that now accounts for 6.92% of non-video game toy sales.

Hasbro's success was quickly copied by Ideal—with Captain Action, by Marx—with Best of the West series, and by many others, but G.I. Joe remained tops until competing company Mego Corporation introduced its first line of action figures in 1972. The company dominated the action-figure business for a good part of the 1970s with superheroes, monsters, TV show and movie characters, plus historic knights and folklore figures. They were riding high with Planet of the Apes and Star Trek series when the film **Star Wars** broke in theaters in the summer of 1977. Mego missed getting the Star Wars license by hours. They had just signed a deal to do Micronauts and were unavailable when the 20th Century Fox representative came calling. The rep took the deal across the street where Bernie Loomis of Kennar purchased the rights on sight because "they were so cheap." No Star Wars toys could be produced until February of 1978, but they were the dominant male action toy by the following Christmas. Mego did well with Micronauts, but the company's days of being the leading action-figure producer were in decline.

By 1981, the second Star Wars film, **The Empire Strikes Back,** had come and gone. Mattel had cranked up the most success-ful male–action-figure line in the company's history—Masters of the Universe. This line was the first to offer a unique action in each figure. For example, Battle Damage He-Man and Skele-tor figures had spring-loaded cylinders in their chests. Each time a miniature sword struck the vulnerable spot, the cylinder would revolve a quarter turn, revealing an additional wound. Hit Ram Man on the head and his whole body collapsed to half of its size. Extendar doubled in height and reach. Moss Man was fuzzy. Sssqueeze had long arms that could tighten like a python around your arms or other figures. Just about every action found in today's figures can be found in the old Master of the Universe series, which sold steadily until 1990.

By 1986, Star Wars, the dominant toy license for eight years, began to run out of steam. Two other major toy sensations had entered the scene—the return of G.I. Joe in a 3¾-inch size in 1982; and Transformers in 1983. Both were from Hasbro. The new G.I. Joe—A Real American Hero revived the larger version that had disappeared in 1978. The smaller size permitted airplanes, tanks and headquarter playsets that weren't possible with the 12-inch scale. There were super toys as part of the line—even a 7-foot aircraft carrier, plus a huge space shuttle and launching tower. Each figure was part of the G.I. Joe line, but unlike the 12-inch G.I. Joes, each character was named. They had a biocard history, mainly written by Marvel Comics writers who did the G.I. Joe comic. The one notable exception was novelist/G.I. Joe fan, Stephen King, who did the biocard copy for "Crystal Ball," a figure that unlike his famous books, didn't sell very well. Small G.I. Joe figures have continued in various forms since their reintroduction.

Changeable robot figures started to appear in Japan in the mid-1960s and showed up in force at the 1982 Japan Toy Fair. Hasbro, which secured the rights for other markets, made plans to introduce them to the United States. The marketing team coined the name "Transformers" and the tag line "Robots in disguise." Mostly Takara figures were used the first year. Transformers have been continually produced by Hasbro ever since, but were not sold in the United States for most of 1991 and 1992.

When Transformers hit the shelves in 1983, it was soon clear Hasbro had introduced its second megahit in as many years. Never before or since has a toy company introduced such classic action-figure lines back-to-back. Though all types of packaging, play testing, kid surveys, contests, and other research is done to predict the success of a toy line, in the end, nobody can really be sure about anything until a supporting TV show or film is released, the commercials have run, and/or the toys are actually on the shelves for sale. Good toy people have a nervous instinct for what will sell two years from the day work begins on a new action-figure line. That is how long it takes to develop the line, present it to retailers, do the tooling, get all the components to East Asia, and ship the goods back to the United States for placement on retail shelves. Some toys premiere in Japan, but most begin in the United States.

Every year, hopes run high that the kids will pick a sleeper as they did with the Mighty Morphin Power Rangers. Presenters of "sure winners" also pray the kids will stick with them one more year. Nobody is ever quite sure what will turn the kids on and make their action-figure line a megahit.

Star Wars, Masters of the Universe, G.I. Joe—A Real American Hero and Transformers dominated the market up until the mid-1980s. Just as Star Wars began to wane, Kenner introduced The Real Ghostbusters, which had a strong run until 1990. The latter half of the 1980s saw the first major importation of Japanese concept toys since Micronauts. Certainly Transformers were created in Japan, but they were given a distinctive U.S. marketing revision. The success of Transformers, however, undoubtedly paved the way for Silverhawks, Voltron and Robotech. There was also a lot of experimentation with interactive action-figure concepts during the period of video game dominance. Toy makers tried Captiain Power, Bravestarr, and Photon. These lines employed various electronic gadgetry to allow kids to shoot at a TV screen to knock off a bad guy or for the bad guy to shoot back at their action figure if they missed. Different lines had figures that could shoot at other figures or had a kid's weapon that could score a hit or cause a figure to react. Sounds like a natural winner? Well, not quite! Kids love interactive video games but yawned at interactive figures. Maybe it was the higher cost, but none of those lines were successful on the interactive feature. The only one that appears in this listing of 1,001 most popular action figures is Captain Power.

The Teenage Mutant Ninja Turtles began capturing the imagination of kids in 1988 and has continued as a core brand for Playmates Toys for over ten years. The total outrageousness of turtles moving at the speed of a ninja caught the kids' fancies. Playmates loaded packaging with equally outrageous puns and plays on stale jokes. The kids loved it. The figures were brightly colored and the painting technology improved over previous action figure lines. Turtles continued at full speed until another import from Japan crept in through the back door.

The Mighty Morphin Power Rangers were shown in Bandai's small Toy Center showroom at the 1993 New York Toy Fair. I sat through the presentation with four toy buyers and was totally bewildered by the presenters transforming this figure and taking parts from two other figures and combining things to make new ones. Everything was so complicated it didn't seem possible parents could ever understand what figures to buy so the kids could tear them apart to make the figures they actually wanted. Most toy buyers must have agreed because of the advance orders for the Mighty Morphin Power Rangers were sparse. Even when the figures hit the shelves not much happened, but once the TV show started airing in September, 1993 the Mighty Morphin Power Rangers started to rewrite toy history. The figures retailed for two dollars more per figure than other lines. Essentially the same figure in five different colors was in such demand Bandai went into 24-hours-a-day production in a losing battle to meet the needs of parents clamoring for Power Rangers and all the Zords that went with them.

Batman had a similar problem in 1989. Toy Biz was a new company back then and wasn't quite prepared for the sensation caused by the first **Batman** film. When demand mushroomed, the company turned to multiple suppliers. The result was three different head variations on the Batman figure further complicating the catch-up situation. Whenever a toy exceeds the expected market, it takes three to eight months to get production and consumer demand back in sync.

Sometimes toy history repeats itself. G.I. Joe has had three totally different successful brand incarnations. Transformers Beast War injected the second major new life into the brand. Star Wars was reintroduced in 1995 and surpassed the initial success it enjoyed from 1978 to 1986 in less than two years. The new Prequel Trilogy guarantees the success of the Star Wars action figure line for at least eight more years beyond the release of Episode I—The Phantom of the Menace on May 19, 1999.

The concept of a male action-figure had survived for thirty-five years. It seems safe to say the idea has clicked and is here to stay. While the first G.I. Joes were detailed, the technology of action-figure production has improved immensely over the years. State-of-the-art manufacturing techniques produce miniature objects of art that function as toys for re-creating scenes from movies or TV shows to small sculptures simply to be admired. Modern action figures are not only artistic, but art anyone can own. As publishers of a monthly action-figure collector's publication since 1991, we have seen the number of adult collectors skyrocket from around 5,000 over twentyfold.

About 20 to 25 thousand different action figure characters have been produced since the first G.I. Joe. They are all listed and most are pictured in **Tomart's Encyclopedia and Price Guide to Action Figure Collectibles** along with all their vehicles, playsets and accessories. Picking the 1,001 most popular action figures ever produced was not an easy task. Other experts might develop a substantilly different list. The criteria I used perhaps are a bit jaded by what collectors have preferred over the years. The retail sales our company has tracked over the last eight years made the job a little more scientific for a small part of action figure history. It doesn't take a genius to realize older lines with only eight to ten figures didn't gain much popularity before they disappeared, but I guarantee that each and every one was some reader's most favored action figure of all time. Individual tastes are different. More difficult from my perspective is picking the most popular from extremely popular lines like 3¾-inch G.I. Joes, where over 400 different figures were made; or choosing the top Star Wars figures from more than 300 choices produced in two batches nearly 20 years apart.

Whether or not your personal taste agrees with mine about the 1,001 action figures selected, one thing is certain: If you were a child between 1964 and 1997, you are going to find some old friends in the pages of this book. Many of these old friends have attained some degree of value, particularly if you managed to preserve them in their original boxes or packages. But value is a fleeting thing with this type of collectible. The ultimate owner cherishes them for the role they have played in his or her own childhood, for their aesthetic appeal, or for the personal admiration of the characters or fantasy they minaturize. Therein lies the foundation for today's success in the male action category—first, kids who go for cool figures with neat play action, and, second, adult collectors who recognize a higher artistic sense while indulging the spirit of youth still inside.

The 1,001 most popular action figures rated high on both scores for the time in which they were created. In some cases, a special story contributed to special notoriety of individual figure creations. For the first half of action-figure history, the kids picked all the most popular figures. Adult-collector influence has increased annually for the last eight to ten years. The result has been better quality and cooler toys for both.

Tom Tumbusch
January 2000
Publisher
Action Figure Digest Magazine

CONTENTS

COMIC ACTION HEROES
(CONTINUED)

BATMAN
ROBIN
THE JOKER
GREEN GOBLIN
SPIDER-MAN
CAPTAIN AMERICA

CONEHEADS   76

BELDAR

COPS (N'CROOKS)   76

A.P.E.S.
SGT. MACE

DAKIN AND DAKIN STYLE FIGURES   77

BIG BOY
SMOKEY BEAR
BUGS BUNNY
WILE E. COYOTE
ROAD RUNNER
TWEETY BIRD
SYLVESTER
BARNEY RUBBLE
FRED FLINTSTONE
PEBBLES
BAMM-BAMM
MIGHTY MOUSE

MICKEY MOUSE
MINNIE MOUSE
GOOFY
DONALD DUCK
YOGI BEAR
SCOOBY DOO
POPEYE
OLIVE OYL
I.A. SUTTON BANANA SPLITS:
   DROOPER THE LION
   BINGO THE BEAR
   FLEAGLE BEAGLE THE DOG
   SNORKY THE ELEPHANT

DARK KNIGHT COLLECTION   81

CRIME ATTACK BATMAN
BRUCE WAYNE
KNOCK-OUT JOKER

DARKWING DUCK   82

DARKWING DUCK
LAUNCHPAD McQUACK

DC COMICS SUPER HEROES   83

SUPERMAN WITH KRYPTONITE RING
WONDER WOMAN
THE PENGUIN, MISSILE-FIRING

FLASH GORDON 96

MEGO:
  FLASH GORDON
  DALE ARDEN
  DR. ZARKOV
  MING, THE MERCILESS
MATTEL:
  FLASH GORDON
  DR. ZARKOV
  THUN, THE LION MAN
  MING, THE MERCILESS
  BEASTMAN

THE FLINTSTONE KIDS 98

CAVEY, JR.
PHILO QUARTZ
WILMA SLAGHOOPLE

GARGOYLES 99

QUICK STRIKE GOLIATH
POWER WING GOLIATH
ELISA MAZA

GHOST RIDER 100

**GHOST RIDER, MOLDED
ONE-PIECE TOY**

12″ GI JOE 101

ACTION SOLDIER
ACTION SAILOR
ACTION MARINE
ACTION PILOT
WEST POINT CADET PHOTO BOX
MILITARY POLICE PHOTO BOX
SKI PATROL
SPECIAL FORCES
GREEN BERET
**ACTION SOLDIER
(AFRICAN-AMERICAN)**
TALKING ACTION SOLDIER
SEA RESCUE
FROGMAN
SHORE PATROL SET
DEEP SEA DIVER
LANDING SIGNAL OFFICER
ANNAPOLIS CADET
TALKING ACTION SAILOR
**COMMUNICATIONS POST &
PONCHO SET**
COMBAT PARATROOPERS
MARINE MEDIC SET
DRESS PARADE SET
AIR CADET IN PHOTO BOX
FIGHTER PILOT IN PHOTO BOX
TANK COMMANDER
JUNGLE FIGHTER
TALKING ACTION MARINE
ASTRONAUT SUIT
TALKING ACTION PILOT

GERMAN STORM TROOPER

JAPANESE IMPERIAL SOLDIER

RUSSIAN INFANTRY MAN

FRENCH RESISTANCE FIGHTER

BRITISH COMMANDO

AUSTRALIAN JUNGLE FIGHTER

ACTION NURSE WITH WHITE BAG

TALKING GI JOE ASTRONAUT

TALKING ADVENTURE TEAM
COMMANDER

CANADIAN MOUNTIE GIFT SET

KUNG FU GRIP LAND ADVENTURER

BLACK ADVENTURER

MOVING EYES LAND COMMANDER

ATOMIC MAN

BULLETMAN

DANGEROUS REMOVAL

SMOKE JUMPER

KARATE

EMERGENCY RESCUE

SECRET AGENT

SUPER JOE

G.I. JOE — A REAL AMERICAN HERO 112

INFANTRY TROOPER—GRUNT

RANGER—STALKER

COMMANDO—SNAKE EYES

MORTAR SOLDIER—SHORT FUZE

COMMUNICATIONS OFFICER—
BREAKER

MACHINE GUNNER—ROCK 'N ROLL

BAZOOKA SOLDIER—ZAP

COUNTER INTELLIGENCE—SCARLETT

LASER RIFLE TROOPER—FLASH

COBRA OFFICER

COBRA COMMANDER

COBRA

MEDIC—DOC

S.E.A.L.—TORPEDO

MINE DETECTOR—TRIPWIRE

MARINE—GUNG-HO

ARCTIC TROOPER—
SNOW JOB

HELICOPTER ASSAULT TROOPER—
AIRBORNE

MAJOR BLUDD

DESTRO

FIRST SERGEANT—DUKE

HEAVY MACHINE GUNNER—
ROADBLOCK

TRACKER—SPIRIT

DOG HANDLER—MUTT & JUNKYARD

COBRA INTELLIGENCE OFFICER—
BARONESS

COBRA SABOTEUR—FIREFLY

COBRA NINJA—STORM SHADOW

FIRST SERGEANT—DUKE

ZARTAN

INFANTRY TROOPER—FOOTLOOSE

SILENT WEAPONS—
QUICK KICK

COVERT OPERATIONS—LADY JAYNE

COMMANDO—SNAKE EYES II

SAILOR—SHIPWRECK

G.I. JOE—A REAL
AMERICAN HERO (CONTINUED)

WARRANT OFFICER—FLINT

COBRA FROGMAN—EELS

**COBRA POLAR ASSAULT—
SNOW SERPENT**

**CRIMSON GUARD COMMANDERS—
TOMAX & XAMOT**

ICEBERG

SCI-FI

ZANDAR

ZARANA

LEATHERNECK

WET-SUIT

HAWK

B.A.T.

DR. MINDBENDER

COBRA COMMANDER

CRYSTAL BALL

BIG BOA

JINX

RAPTOR

CROC MASTER

GUNG-HO (DRESS UNIFORM)

BLOCKER

AVALANCHE

MAVERICK

DODGER

SHOCKWAVE

BUDO

VOLTAR

SPEARHEAD & MAX

STORM SHADOW II

HOODED COBRA COMMANDER

SGT. SLAUGHTER

THE FRIDGE

SERPENTOR

SNAKE EYES III

T.A.R.G.A.T.

GNAWGAHYDE

LIFELINE

RED STAR

DESTRO II

WILD BILL

EFFECTS

LOBOTOMAXX

PREDACON

CARCASS

30TH ANNIVERSARY ACTION SOLDIER

30TH ANNIVERSARY ACTION SAILOR

30TH ANNIVERSARY ACTION MARINE

30TH ANNIVERSARY ACTION PILOT

**30TH ANNIVERSARY ORIGINAL
ACTION TEAM**

HALL OF FAME
12″ G.I. JOES   128

TARGET DUKE

COBRA COMMANDER

SNAKE-EYES

DESTRO

STORM SHADOW

MARVEL SUPER HEROES
SECRET WARS   168

CAPTAIN AMERICA

SPIDER-MAN

IRON MAN

WOLVERINE

DOCTOR OCTOPUS

SPIDER-MAN, BLACK COSTUME

HOBGOBLIN

ICEMAN

CONSTRICTOR

ELECTRO

M*A*S*H   170

HAWKEYE

HOT LIPS

HAWKEYE

THE MASK   172

WILD WOLF MASK

BELLY BUSTIN' MASK

CHOMPIN' MILO

TALKING MASK

MASTERS OF
THE UNIVERSE   174

BATTLE ARMOR HE-MAN

BATTLE ARMOR SKELETOR

BEAST MAN

BUZZ-OFF

BUZZ-SAW HORDAK

DRAGSTOR

EVIL-LYN

EXTENDAR

FAKER

FISTO

HE-MAN

HURRICANE HORDAK

JITSU

KING HISS

KING RANDOR

KOBRA KHAN

LEECH

MAN-AT-ARMS

MAN-E-FACES

MANTENNA

MEKANECK

MER-MAN

MOSQUITOR

MOSS MAN

NINJOR

ORKO

PRINCE ADAM

RAM MAN

ROKKON

ROTAR

SAUROD

SCARE GLOW

SKELETOR

SNAKE FACE

SNOUT SPOUT

STEELWILL
MOON STRYKER
CONDOR
WINDHAMMER

THE SIMPSONS   235

BART
HOMER
MARGE
LISA
MAGGIE
NELSON
BARTMAN

THE SIX MILLION DOLLAR MAN   237

COLONEL STEVE AUSTIN
OSCAR GOLDMAN

SPAWN   238

SPAWN (CLAMSHELL WITH COMIC)
MEDIEVAL SPAWN
CLOWN
ANGELA
PILOT SPAWN
MALEBOLGIA
COSMIC ANGELA
SPAWN II
VIOLATOR II

VERTEBREAKER
FUTURE SPAWN, BLACK/RED
CY-GOR
THE MAXX
SHE-SPAWN
EXO-SKELETON SPAWN
ZOMBIE SPAWN
SPAWN III
MOVIE CLOWN
MANGA MEDIEVAL SPAWN
MANGA VIOLATOR
MANGA ANGELA

SPIDER-MAN   242

SPIDER-MAN, WEB SHOOTER
DR. OCTOPUS
CARNAGE
VENOM
PETER PARKER
KINGPIN
THE LIZARD
10″ SUPER POSEABLE SPIDER-MAN
GREEN GOBLIN
SHOCKER
SCORPION
THE RHINO
CAPTAIN AMERICA
ELECTRO

STAR TREK   246

CAPT. KIRK
MR. SPOCK
NEPTUNIAN
ANDORIAN
THE ROMULAN
TALOS
ARCTURIAN
BETELGEUSIAN
MEGARITE
REGELLIAN
ZARANITE
LIEUTENANT TASHA YAR (GALOOB)
CAPTAIN JEAN-LUC PICARD
LIEUTENANT COMMANDER DATA
DATHON
LIEUTENANT COMMANDER GEORDI LA FORGE
ROMULAN
VORGON
LIEUTENANT WORF
COUNSELOR DEANNA TROI
AMBASSADOR SPOCK
CAPTAIN SCOTT
BENZITE
ESOQQ
Q IN JUDGE'S ROBE
CAPTAIN JEAN-LUC PICARD (AS DIXON HILL)
AMBASSADOR K'EHLEYR
AMBASSADOR SAREK

BORG
THE HUNTER OF TOSK
QUARK
COMMANDER GUL DUKAT
ROM WITH NOG
KIRK
SULU
SPOCK
KHAN
DR. McCOY
LIEUTENANT UHURA
LIEUTENANT SAAVIK
KIRK IN SPACE SUIT
CHEKOV
GUINAN
ILIA PROBE
SWARM ALIEN

STAR WARS   259

LUKE SKYWALKER
PRINCESS LEIA ORGANA
CHEWBACCA
ARTOO-DETOO (R2-D2)
HAN SOLO
SEE-THREEPIO (C-3PO)
STORMTROOPER
DARTH VADER
BEN KENOBI
JAWA
SANDPEOPLE

SNAGGLETOOTH

WALRUS MAN

BOBA FETT

LUKE SKYWALKER (BESPIN FATIGUES)

IMPERIAL STORMTROOPER
(HOTH GEAR)

LEIA (BESPIN)

IG-88

YODA

UGNAUGHT

LOBAT

HAN SOLO BESPIN

IMPERIAL TIE FIGHTER PILOT

HAN SOLO IN CARBONITE

EMPEROR'S ROYAL GUARD

LUKE SKYWALKER JEDI KNIGHT

PRINCESS LEIA ORGANA
(BOUSHH DISGUISE)

SQUID HEAD

ADMIRAL ACKBAR

TEEBO

ARTOO-DETOO WITH
POP-UP LIGHTSABER

ANAKIN SKYWALKER

LUKE SKYWALKER IN IMPERIAL
STORMTROOPER OUTFIT

IMPERIAL GUNNER

A-WING PILOT

AMANAMAN

YAK FACE

SY SNOOTLES AND THE REBO BAND

STAR WARS   271

DARTH VADER

CHEWBACCA

BEN OBI-WAN KENOBI

C-3PO

BOBA FETT

LUKE JEDI KNIGHT

HAN SOLO IN CARBONITE BLOCK

LUKE IN STORMTROOPER DISGUISE

DEATH STAR GUNNER

CANTINA BAND MEMBER

EMPEROR PALPATINE

GRAND MOFF TARKIN

GARINDAN

GAMORREAN GUARD

YAK FACE

CEREMONIAL LUKE

GENERAL LANDO CALRISSIAN

DARTH VADER WITH
REMOVABLE HELMET

BESPIN LUKE

EWOK CEREMONIAL PRINCESS LEIA

R2-D2 WITH DATA LINK

SUPERMAN   276

SUPERMAN

Tomart's Encyclopedia of

Action Figures

THE A-TEAM

GALOOB, 1984–85
SIZE: 6"
VALUE RANGE: $12–$20

MR. T AS B.A. BARACUS

Stephen J. Cannell's hit TV show **The A-Team** for the ABC television network made a star out of a bejeweled, Mohawk-wearing Mr. T. As Sergeant Bosco "B.A." (Bad Attitude) Baracus, T was everyone's favorite team member. The B.A. Baracus 6" figure

was fully poseable and came with a "full set of action accessories" and weapons, like bolt cutters, an M-16 rifle and shoulder strap, 6 other action tools in a toolbox, and of course, golden neck chains. The A-Team action figure line included both 6" and 3¾" figures, as well as different adventure sets, vehicles and a combat headquarters set.

MR. T AS B.A. BARACUS

ACTION JACKSON

MEGO, 1974
SIZE: 8"
VALUE RANGE: $12–$40

ACTION JACKSON

"Action Jackson is his name! Bold adventure is his game! Think of what you want to be, then call on me— Action Jackson!" Action Jackson was

Mego's generic action hero, most likely designed to compete with G.I. Joe. The 8", jump-suited male figure came with stick-on tattoos and "mod styled" hair, but what baby boomers remember most about Action Jackson is the catchy song from the television commercials. Much like G.I.

ACTION JACKSON

Joe, Action Jackson was an all-around hero. Accessory kits available included a strap-on helicopter with propeller action, a secret agent outfit, a fire rescue pack, and many more action- and military-related equipment. A feature film based on the action figure was made in the mid-1980s featuring Carl Weathers, but it was not greeted with much success.

ADVANCED DUNGEONS AND DRAGONS

LJN, 1983–84
SIZE: 5"
VALUE RANGE: $8–$20

STRONGHEART
WARDUKE

These finely detailed action figures came from a Saturday-morning TV cartoon series, which in turn was based on the popular TSR medieval fantasy role-playing game. The "good" action figures were packaged on yellow-gold rectangular cards, and the "evil" characters were on dark blue cards. "Strongheart," a good paladin or protector,

STRONGHEART

WARDUKE

came with a sword. The detailing on Strongheart, as well as Warduke and the other figures in this line, is some of the most intricate ever done on an action figure. His regal blue and silver armor and blue cape are stunningly rendered, and he possesses a dashing, mus-

tachioed face. Warduke, an evil fighter, came with a black sword and a shield with a demonic face engraved on it. He has no visible counte-nance, but a dark, masked face with glowing red eyes and batwinglike ears.

ADVENTURES OF INDIANA JONES

KENNER, 1982–83
SIZE: 3¾"

INDIANA JONES
MARION RAVENWOOD

The success of **Raiders of the Lost Ark** took everyone by surprise. Fortunately, Kenner had a relationship with Lucasfilm Ltd. because of the extraordinary success of the **Star Wars** toys, and produced "The Adventures of Indiana Jones" line of action figures and accessories. But the figures were underproduced in the first year, marketing was spotty, and the line never made a big impact on toy buyers. Nevertheless, the demand for the Indiana Jones figures far exceeded the supply, and eventually interest in the figures waned when shop-

INDIANA JONES/$135

pers couldn't find the major character toys in the stores. Two of the secondary figures were still on close-out shelves a full ten years after their release.

Only four figures were released the first year. The fully articulated nine action figures from this line are extremely valuable. The extremely rare Marion Ravenwood figure came wearing a white lace fabric dress, and was packaged with her pet monkey.

MARION RAVENWOOD/$200

She was released sparingly in the second year of the line and was shipped only one per case. There were only two or three Indiana Jones figures per case while new secondary characters dominated the assortment. The Indiana figure came dressed in his trademark field outfit and had a "fast-draw action arm" that allowed him to draw his gun or crack his whip (both also included).

DISNEY'S ALADDIN

MATTEL, 1992–95
SIZE: 5"
VALUE RANGE: $2–$7

ALADDIN
GENIE
JASMINE AND RAJAH

Mattel's delightful action figures based on the Disney animated feature **Aladdin** were a smash in the stores. Here were nicely detailed fun figures that were perfect for kids who weren't interested in more violently themed toys. The figures looked very much like their movie counterparts and came on colorful cards with scenes from the movie. Aladdin was packaged with his

ALADDIN

GENIE

magic carpet, the irrepressible Genie came with his magic lamp, and Princess Jasmine came with her pet tiger, Rajah. The Genie's head could be flipped over to give him a different look. Other versions in this line included "battle action" figures, including a downright frightening Evil Genie Jafar. The Genie was also available as the "Frenchman" Genie, the "Top Hat and Tails"

JASMINE AND RAJAH

Genie, and the "Baseball Player" Genie. These figures perfectly reflected the great whimsy of the Genie's character.

ALIEN, ALIENS

KENNER, 1979, 1992–95
SIZE: 5"

ALIEN (18")
LT. ELLEN RIPLEY
SCORPION ALIEN
ALIEN QUEEN
ALIEN ARACHNID
KING ALIEN

One of the most disturbing action figures to make it to toy store shelves, albeit briefly, was a replica of one of contemporary cinema's most frightening monsters, the savage, bloodthirsty Alien of Ridley Scott's space-horror classic film, **Alien**. This 18" authentic reproduction of the movie creature had retracting inner teeth, spines on its back, mechanically operated fanged jaws, a moveable tail, and glow-in-the-dark head-paint. He was fully articulated at the hips and shoulders, and his arms were spring-loaded to "crush its

ALIEN/$200–$800

victims." A spectacular action figure, the Alien (and other **Alien** merchandise) was deemed too frightening and was removed from toy store shelves almost as quickly as it had appeared. When one looks at action figures of the late 1990s, it is difficult to understand why this figure was judged to be so horrifying.

LT. ELLEN RIPLEY/$8–$18

SCORPION ALIEN/$8–$18

The sequels to **Alien** were huge commercial successes and cemented Sigourney Weaver's Ripley as one of the first true female action heroes.

continued on page 48

45

The Cancelled Alien 4½" Line

Legend was the first Alien action figure to be released by Kenner. This classic 18" figure offered in conjunction with the film's release was perceived—by parents—to be so frightful and distressing to kids that it became a media event. The clamor was so loud that shipments were stopped. Of course, the kids thought it was cool. Time has proven that many "kids" wanted this sensational toy, and they have driven the price to more than $600 when found in like-new, boxed condition. Even completely loose figures are often seen priced at more than $300.

What very few people ever knew was the simultaneous loss of the 4½" line. This smaller line did not appear in any Kenner catalog. The 18" figure was introduced at Toy Fair in Spring 1979, with a plan to add the 4½" line in time for Christmas. There are five known figures in

RIPLEY (IN DECK UNIFORM), ASH AND DALLAS

SPACESUIT RIPLEY (REAR VIEW), ALIEN WITH TONGUE RETRACTED (SIDE VIEW)

RIPLEY (3/4 VIEW), ALIEN WITH TONGUE EXTENDED

FRONT VIEW OF RIPLEY (W/HELMET REMOVED) AND THE ALIEN

the line. Perhaps some accessory vehicles or even a play set were also planned.

The 4½" Alien figure is all black, with paint details on the head only. The white node atop the skull slides a white forked tongue between the exposed pointed teeth.

There were two Ripley figures: one in her deck uniform and a second in a highly detailed Moebius-designed spacesuit. The helmet on this suit was removable, and it could be argued that this figure was more interesting from the back rather than the front.

The other two known figures from this series are crewmembers Dallas and Ash, both in their deck uniforms. Odds are, the remaining crewmembers and at least one more version of the Alien were produced in this cancelled Alien line. Collectors surely regret their inability to ever purchase any of these figures.

ALIEN QUEEN/$8–$18

ALIEN ARACHNID/$15–$30

KING ALIEN/$8–$18

The success of the movie also insured that the alien creature(s) would enter the ranks of classic movie monsters such as Godzilla. Such commercial and cultural appeal demanded that the **Alien** action figure line be revived, and in 1992, Kenner did just that.

The Ripley figure is dressed in the character's trademark T-shirt and carries a large-barreled gun with real "turbo torch action." Ripley also comes with Marine decals

that can be placed on the weapon or on the figure's clothing. The card features a full-color painting of Ripley and an oversized Alien whose claw seems to be reaching for the figure in the plastic bubble. It's a clever packaging effect. Ripley and the Scorpion Alien also came with a free **Dark Horse** mini-comic book inside.

While Ripley is a fine action figure, the imagination and artistry of the figures' designers

really shine through in the Alien figures. Loosely based on the creatures from the films, these figures are absolutely fantastic and terrifying at the same time. They are exquisitely detailed, and their two-color (black and a dark color) composition is very effective. The Scorpion Alien looks the most like the creature in the first film and even comes with a "face hugger." It will break apart or "explode" when hit. The Alien Queen comes with a "deadly chest hatchling," double jaws and a spiked tail.

The Fourth Series Alien Arachnid and King Alien have slightly different packaging. Alien Arachnid was packaged against a card of a red Alien battling a silver Alien, while the Alien leader or King shows the same silver Alien fighting a green Alien creature. Alien Arachnid has "venom spray action," and the King has "crushing" grip and action spray as well.

ALIENS VS. PREDATOR

KENNER, 1993
SIZE: 6"
VALUE RANGE: $15–$25

**WARRIOR ALIEN/
RENEGADE PREDATOR
(2-PACK)**

This cross between two science-fiction movie creatures was marketed as "the ultimate battle between Beast and Hunter." These finely detailed creatures were the only action figures sold under the "Aliens vs. Predator" theme in a two-pack.

**WARRIOR ALIEN/
RENEGADE PREDATOR**

THE AMAZING SPIDER-MAN

MEGO, 1978–81
SIZE: 12"
VALUE RANGE: $25–$60

There are several Spider-Man action figures mentioned in this book, but this figure was not part of a larger series. This Spider-Man was a 12½" poseable figure dressed in a cloth costume. He was originally designed to also have magnetic hands and feet, but those plans were scrapped before production.

THE AMAZING SPIDER-MAN

ANNIE

KNICKERBOCKER, 1982
SIZE: 6"
VALUE RANGE: $5–$20

ANNIE W/LOCKET
MOLLY

This line was created to coincide with the release of the movie musical **Annie**, which was based on the hit Broadway show of the same name. Although the **Little Orphan Annie** comic strip dates back to August 5, 1924, no action figures of the feisty foundling were made until the 1980s.

Both Annie and Molly, Annie's best friend,

ANNIE W/LOCKET

MOLLY

are about 6" tall and were sold in blue cardboard window boxes with a photo of the main cast members printed on the front. The first Annie comes with the little one's trademark red curly hair and dress and a child-sized

locket "for you!" Molly is dressed in a turquoise jumper and flowered blouse, with long brown hair. Both girls are smiling. They were terrific toys for small children (once the locket was removed by a responsible parent).

BATMAN

MEGO, 1979
SIZE: 12½"
VALUE RANGE: $40–$120

BATMAN

TOY BIZ, 1989
SIZE: 8½"
VALUE RANGE: $8–$20

BATMAN (3 VARIATIONS)

This poseable 12½" Batman from Mego was based on the comic book. He wears a cloth costume and plastic cape. He comes in a window box with a comic-book drawing and is pictured with Robin and Superman. Robin was also available in the series. He is not as dark and heavily muscular as the figures based on the **Batman** movie but is very close to the earlier comic-book hero.

The Batman series by Toy Biz was released in conjunction with the first **Batman** movie, starring Michael Keaton. The

BATMAN

BATMAN (PAINTED FACE)

BATMAN (SQUARE JAW)

BATMAN (ROUND JAW)

small start-up company was hard-pressed to meet demand, so figures were sent to outside companies for the tooling of extra molds. This resulted in three Batman action figure variations—one had a painted face, one had a square jaw, and one,

a round jaw. Most collectors want all three versions. This Batman's action features include a retractable mechanism in his utility belt that actually allows him to climb the Bat-rope, also included. The figure also comes with a Bat-a-rang and a speargun.

BATMAN—THE ANIMATED SERIES

SIZE: 5"
VALUE RANGE: $10–$60

COMBAT BELT BATMAN
TURBOJET BATMAN
ROBIN
THE RIDDLER
THE PENGUIN
BRUCE WAYNE
THE JOKER
MAN-BAT
CATWOMAN
SCARECROW
MR. FREEZE
POISON IVY
HARLEY QUINN

COMBAT BELT BATMAN

TURBOJET BATMAN

ROBIN

THE RIDDLER

The **Batman** films sparked a renewed interest in all things Batman, and in 1992, Warner Bros.' **Batman—The Animated Series** first aired on TV. While the early Tim Burton films **Batman** and **Batman Returns** were hailed for their Gothic design and ambience, the animated TV series gave birth to a new technique in animation using black backgrounds that would eventually be dubbed "Dark Deco." The TV series included all the popular characters, revamped the classic characters, and invented some new ones. The most significant change was the transformation of the

THE PENGUIN

BRUCE WAYNE

THE JOKER

MAN-BAT

CATWOMAN

SCARECROW

Dick Grayson/Robin character in the "new" 1990s costume, resulting in a hipper, more adult representation of the "Boy Wonder."

Even though most of the line was dominated by similar Batman figures, both the art deco packaging and the action figures

were painted to reflect a brighter mood. The painting on the packaging of all the figures is dark and iconographic, with

MR. FREEZE

POISON IVY

HARLEY QUINN

Batman's head show-ing only white eyes on a black mask against a red moon.

The villains include fanciful accessories

like Catwoman's pet cat and the Scare-crow's scarecrow. The Joker has a laughing gas spray gun, and Poison Ivy is pack-

aged with a snapping Venus's-flytrap. All fig-ures are painted with an eye toward detail and faithfulness to the TV series.

BATMAN—MASK OF THE PHANTASM

KENNER, 1994
SIZE: 5"
VALUE RANGE: $8–$28

PHANTASM

Batman—Mask of the Phantasm was a full-

length animated fea-ture based on **Bat-man—The Animated Series**. The animation was high-quality "Dark Deco." Warner Bros. had originally planned to release this film direct to video, but once they saw the fin-

ished product, they decided it was good enough to debut on the big screen. Many critics agreed, and some even consider this feature to be closer to the dark Tim Burton versions and far superior to the later live-action films in terms of story, design, and characters.

The villain of the movie is Phantasm, a masked vigilante who murders criminals. The action figure hides behind a frightening skeleton mask and slashing sickle when up against Batman. Phantasm features "Chopping Arm Action" and comes with a Blade weapon that fits in her right hand and a Laser weapon that fits into her left hand. Her Cowl and Mask can be removed to reveal that she is actually a very feminine brunette. Typically shy of female action figures, Kenner only packaged one Phantasm per case. Ironically, this figure dominates collectors' interest in this particular **Batman** line.

PHANTASM

BATMAN RETURNS

KENNER, 1992–94
SIZE: 5"
VALUE RANGE: $10–$40

CATWOMAN
THE PENGUIN

This series was an extension of Kenner's **Dark Knight Collection.** It was renamed and expanded to coincide with the 1992 release of the

CATWOMAN

THE PENGUIN

second Batman movie, **Batman Returns**.

Catwoman is modeled after Michelle Pfeiffer's character and comes with a whip and a laser gun. The package art is superb and shows, in detail, the Catwoman costume that Pfeiffer donned so memorably in the movie.

The Penguin is a bit more comical than the dark, sociopathic villain of the movie because Kenner reused tooling previously created for its Super Powers collection. This Penguin is fatter and resembles the comic-book Penguin more than Danny DeVito. He comes with a blast-off umbrella launcher.

BATTLESTAR GALACTICA

MATTEL, 1978
SIZE: 5″
VALUE RANGE: $10–$65

LT. STARBUCK
CYLON COMMANDER

Battlestar Galactica, a 1978 science-fiction television series, had a much darker vision of the future than **Star Trek** ever did, and it is perhaps for this reason that the show did not succeed with the millions of **Trek** fans it hoped to reach. Since the growth of cable, however, **Battlestar**

LT. STARBUCK

Galactica has been airing in reruns and has developed a large following. Trendmasters acquired the license in 1996 and began plans for a new line of figures.

CYLON COMMANDER

The story is quite compelling. The 12 Colonies of Man are all but wiped out by a cybernetic race called Cylons. Commander Adama and the Battlestar Galactica lead a

56

human fleet of survivors in search of a mythical planet called Earth, battling Cylons all the way.

The human figures are nothing special. Lieutenant Starbuck is a basic human male action figure with a weapon. The Cylon Commander, however, is a menacing, gold-metallic figure. He was available as a mail-in premium as well as on the toy store shelves.

BEST OF THE WEST

MARX, 1965–76,
1993 (REPRODUCTIONS)
SIZE: 12"
VALUE RANGE: $25–$75

JOHNNY WEST
JANE WEST
JOSIE WEST
GERONIMO

This group is a good example of the early action figure lines.

These figures are of remarkable quality and reflect a care in concept and design that one rarely sees any longer.

Each 9" to 12" figure came in a box with a drawing showing the enclosed toy. Packaging varied widely over the years. In the first year of production, Johnny was the "action cowboy" and Jane was an "action cowgirl." Josie was produced in 1973, and her box had more of a mod look. She was billed as a "movable cowgirl." Geronimo, a "movable Apache Indian" was first manufactured in 1967, and his box showed a photograph of the toy

JOHNNY WEST

JANE WEST

JOSIE WEST

GERONIMO

with a Western art-work theme in the background.

Complete figures with all the accessories,

and there were many, are the most prized. Each figure had at least 12 pieces of Western gear, molded soft plastic clothing

and personal acces-sories. Geronimo has a bow, a shield and several other "Indian" accessories.

BIG JIM

MATTEL, 1973–76
SIZE: 9"
VALUE RANGE: $20–$45

BIG JIM

Big Jim was Mattel's attempt to take some of the action away from G.I. Joe. Big Jim was an athlete, adven-turer, and sportsman, not a military man.

Mattel called Jim a "sports hero" and promised "a world of adventure and sports fun." Big Jim could do everything from skin diving to skiing in the Olympics to exploring the Arctic, but he couldn't take the mar-ket away from G.I. Joe. The basic Big Jim figure is 12" and has

BIG JIM

continued on page 60

58

The Steel Glove:
An Action Figure Comes to Life

Big Jim was introduced and marketed as an athletic, clean cut superstar who excelled in every type of sport and recreational activity. So was his image from 1973 to 1975 when sales were sliding dramatically. Big Jim was in need of a completely new image if he was to survive. Mattel came up with a good-guy vs. bad-guy theme for 1976 called Big Jim's P.A.C.K. (Professional Agents Crime Killers)

Large-order toy buyers were convinced that this new Big Jim deserved a second chance and Toy Fair '75 was a success for Joel Rubenstein. His sales force got excited over all the life-size accessories, especially the Dr. Steel glove the Boys' Toy leader had made for himself.

hand into his trench coat pocket. He calmly slid his hand into the steel glove. As the men grew nearer, he slowly removed his gloved hand from the pocket and raised it to his chin. The steel hand flashed reflections from the station lights and the men froze in their

Awaiting a late train home after a long, grueling toy fair, Rubenstein noticed three thug-like characters approaching him on the platform. Not knowing what their intentions might be, he slipped his

tracks. The train arrived not a moment too soon. Who knows what might have happened but for the life-size Dr. Steel glove?

flexing biceps and swinging arms and a nicely sculpted, buff torso.

Nice guys don't always finish last. Bad Guys were introduced along with Big Jim's

P.A.C.K. in 1973 and the line had a major resurgence.

THE BIONIC WOMAN

KENNER, 1976–78
SIZE: 12"
VALUE RANGE: $25–$75

JAIME SOMMERS

"Beauty and Fashion, **plus** Bionic Action!" is how Jaime Sommers, the Bionic Woman, was described in the 1976 Kenner catalogue. **The Bionic Woman**, a spin-off of **The Six Million Dollar Man**, featured Lindsay Wagner as a bionically enhanced superwoman. Her action figure is 12" tall and fully articulated. It has

rooted hair, an attractive face modeled on Lindsay Wagner, and bionic ears that make a "ping" sound when her head is turned. The Bionic Woman was closer to Barbie than an action figure, and her accessories emphasized this. Indeed, she did have a bionic computer, but it was part of a "Bionic Beauty Salon" where she could prepare for "an evening out with Steve Austin." Most of her accessories were very "Barbie-like"—a geodesic "Dome House,"

JAIME SOMMERS

a "Bubblin' Bath 'n Shower," and a sports car. Two lines of "Designer Collection Fashions" and a line of "Designer Budget Fashions" were also available for the figure.

THE BLACK HOLE

MEGO, 1980
SIZE: 3¾"
VALUE RANGE: $20–$75

V.I.N. CENT.
MAXMILLIAN

The Black Hole was a 1979 live-action Disney movie about a spaceship, **The Palomino**, stationed at a black hole, which discovers a long-lost space vessel, the **Cygnus,** operated by a human, the evil Dr. Hans Reinhardt, and his army of robots. Ten small fig-

V.I.N. CENT.

MAXMILLIAN

ures were made by Mego, and the robots are generally more difficult to find. V.I.N. CENT. is a "good" robot who bears more than a slight resem-

blance to R2D2, and Maxmillian is an "evil" robot who assists Dr. Reinhardt. Both of these robot figures are highly prized by collectors.

BLACKSTAR

GALOOB, 1983–85
SIZE: 6"
VALUE RANGE: $8–$30

BLACKSTAR AND TROBBIT

The **Blackstar** action figures are based on the Filmation animated television series, **Blackstar**. The series told the adventures of

John Blackstar, an astronaut—actually, a sword-carrying astronaut in a cape wearing briefs and a wrestling belt—lost in a strange universe. Blackstar and Trobbit come packaged together. Blackstar is a well-muscled action

BLACKSTAR AND TROBBIT

figure, with bulging biceps, triceps, quads, pecs—you name it. In fact, he looks a lot like He-man, but he looks a lot happier. He comes with a glow-in-the-dark Star Sword. His cute little sidekick, Balkar, is king of the Trobbits, and has the grumpy/adorable look of a Disney-esque character. Both figures have very nicely detailed, expressive faces.

BONANZA

AMERICAN CHARACTER, 1966
SIZE: 12"
VALUE RANGE: $40–$100

LITTLE JOE
HOSS
BEN

The **Bonanza** action figures, or "full action men" as they were called by American Character, are exquisite. These early action figures were based on the legendary NBC drama of the Cartwright family. They are more articulated than most figures and bear close resemblance to the characters they portray.

All three 12" figures come with "authentic accessories." Little Joe, Ben, and Hoss come with a vest, and Little Joe and Ben wear bandanas. The weapons are beautifully detailed, and the costumes are authentic right down to the spur straps. The men are fully poseable—every joint bends and turns. Each comes in cardboard box with full-color drawings of the enclosed figure, along with a color painting of the other available figures. Inside is a brochure describing the entire **Bonanza** line, including their horses and the Bonanza 4-in-1 Wagon, which included two workhorses.

LITTLE JOE, HOSS & BEN

BUCK ROGERS

MEGO, 1979
SIZE: 3¾"
VALUE RANGE: $9–$60

BUCK ROGERS
WILMA DEERING
TWIKI

The live-action television series, **Buck Rogers in the 25th Century,** first aired in September 1979 and was an update of the original 1930s character. In this version, an accident during the test flight of Ranger 3, a NASA deep-space probe, freezes Captain William Anthony "Buck" Rogers in suspended animation. Knocked far from its planned trajectory, Ranger 3 returns to Earth 500 years later. Gil Gerard played Captain William "Buck" Rogers, Erin Gray was

BUCK ROGERS

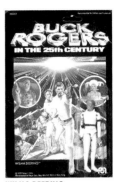

WILMA DEERING

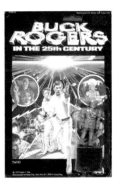

TWIKI

Colonel Wilma Deering and Mel Blanc provided the voice of Buck's little robot pal, Twiki.

The human figures are fully articulated, dressed in white space suits. No action accessories are included. Twiki bears a close resemblance to C3PO, but is nowhere as detailed or animated as the famous **Star Wars** droid. All three are packaged on a card with terrific renderings of action scenes from the show.

BUCKY O'HARE

HASBRO, 1991
SIZE: 5"
VALUE RANGE: $2–$9

#1 BUCKY O'HARE
#10 STORM TOAD
TROOPER

The Space Adventures of Bucky O'Hare was a quirky comic book and animated television series. This series was designed to compete with the highly successful Teenage Mutant Ninja Turtles. The action figures do a great job of conveying the attitude and

#1 BUCKY O'HARE

#10 STORM TOAD TROOPER

humor of the characters deeply embroiled in the "Toad Wars." Bucky O'Hare is the hero of S.P.A.C.E. (Sentient Protoplasm Against Colonial Encroachment) in the toad/mammal con-

flicts, and the Storm Toad Trooper is one of many creatures intent on shooting the "mammals and not other toads." It's more than a little tongue-in-cheek nod to **Star Wars**.

BUTCH AND SUNDANCE: THE EARLY YEARS

KENNER, 1979
SIZE: 3¾"
VALUE RANGE: $5–$30

BUTCH CASSIDY
THE SUNDANCE KID

These figures were based on the title

characters from the 1979 movie **Butch and Sundance: The Early Days**. The film starred William Katt as the Sundance Kid, and Tom Berenger as Butch Cassidy. While not as successful as the original **Butch Cassidy and the**

Sundance Kid, this film was a good adventure story and received an Academy Award nomination for Best Costume design.

Both figures can draw their pistols by using special button-activated "fast-draw

continued on page 66

Bucky O'Hare: Kids Had Other Ideas

In 1991, Hasbro selected **Bucky O'Hare** to increase its market share against the then high-riding **Teenage Mutant Ninja Turtles** from Playmates. It had all the elements of the **Turtles**—size, bright colors, outrageous characters, loads of neat accessories, humor and similar packaging. The company was confident in its product and had the second year of figures designed and ready to go.

Kids had other ideas. For whatever reasons, the line didn't sell well enough for retailers to buy into a second-year production. The poor sales illustrate how unpredictable and quirky a toy line can be. There are few real trends. Kids go for it or they don't.

Second year figures are usually better than first year figures because the designers actually have seen the animation and get a better feel for

JENNY

the character. First year figures are designed from drawings and written descriptions. The required advance time limits product development to occur using a scant amount of information. If the company waits until after a new property is a hit, the demand could easily be missed entirely, as was proven in the case of the fantastically successful film **E.T.**

PIT STOP PETE

RUMBLE BEE

TOTAL TERROR TOAD

action" and are dressed in Western gear. Butch wears his trademark bowler, and Sundance wears a broader-brimmed black hat. The figures are packaged in a plastic bubble on full-color cards with a wood-grain background and a Western-style cameo picture of the character. The typeface is the

BUTCH CASSIDY

THE SUNDANCE KID

same kind as seen on old-fashioned "wanted" posters from the Old West.

CAPTAIN ACTION

IDEAL, 1966–68
SIZE: 12"
VALUE RANGE:
$150–$1,200

CAPTAIN ACTION
ACTION BOY
DR. EVIL

Today we think in terms of collecting a whole set of action figures. Back in 1966, "action figures" were a new concept patterned after the success of Barbie. G.I. Joe was a smash in 1964 with a parade of different uniforms and barrage of carded accessories. It was natural for the Ideal Toy Company to stick close to this proven success formula when they conceived Captain Action, a unique character who could turn into just about any popular super hero. The transformation was accomplished by changing uniforms and adding a rubber mask created true to the likeness of the character being impersonated. The first year, Captain Action could become Superman, Batman, Aquaman, Steve Canyon, The Lone

CAPTAIN ACTION

ACTION BOY

DR. EVIL

Ranger, Flash Gordon, Captain America or The Phantom. Licensing so many different character properties for a single line was a feat—even back then—and would probably be impossible today.

Although in the 1960s Captain Action didn't enjoy the success of G.I. Joe, many consider him to be a stronger commodity in today's market because of the limited supply that lack of popularity generated. Captain Action's costumes consisted of

many small accessories, which have been lost or destroyed over the years, adding to the value and nostalgia associated with the figure.

Super hero characters have turned out to be some of the most successful and steady-selling action figures. This first offering was really a collection of comic-book heroes. In 1966, when the toys were being developed, no one at Ideal could have predicted the tremendous boost

the line would receive from the Adam West **Batman** TV show, which began airing shortly thereafter. **Batman** took the nation by storm and triggered a "pop" revival of virtually every cartoon character. These characters remained popular until the mid-1970s.

The second year, Flash Gordon and Sgt. Fury were dropped in favor of Spider-Man, The Green Hornet, Tonto and Buck Rogers. "Video-Matic" flasher rings were added to the costume and accessory boxes. Action Boy was introduced along with costume ensembles for Robin, Aqualad and Superboy. Six accessory packs were added to give Captain additional powers when he just wanted to be himself, plus a turbo rocket-firing Silver Streak Amphibian vehicle.

Dr. Evil was the next major addition. The standard figure came in a blue smock-type suit with sandals, disguise mask, medallion, and laser gun. He was available boxed or in a gift set that included an extra lab coat, Oriental disguise, and five intergalaxial deadly weapons: a reducer, hypnotic eye, ionized hypo, laser ray gun and thought control helmet. This gift set is commonly referred to as "Dr. Evil's laboratory," even though no such name appears on the box. There was a playset/carrying case depicting several Captain Action and super hero characters on the exterior, but nonetheless called Dr. Evil's Sanctuary.

Dr. Evil also ushered in a new style of figure packaging—the photo box. Instead of the artwork that graced the earlier versions, these new boxes pictured the actual figures in action poses amid a diorama setting. On his box, Captain Action is shown standing triumphant over a prone Dr. Evil, while their roles are reversed on the evil Doctor's box.

Captain Action was discontinued by 1970. The Ideal Toy Company was eventually purchased by CBS as a diversification move. It was reorganized with other CBS properties and lost its identity, even though CBS later sold all its toy company properties. Ideal continues to operate out of England, selling toys throughout Europe. Mego came along in 1972 and dominated the super hero figure category for the next 12 years.

CAPTAIN POWER

MATTEL, 1987
SIZE: 5"
VALUE RANGE: $4–$10

**CAPTAIN POWER
LORD DREAD**

The Captain Power and the Soldiers of the Future line was designed to interact with the **Captain Power** television show, and as the packaging promised, "it really works!" Encoded signals broadcast with the TV show determined hits, or "destroyed" the toys. Videotapes were sold so that children could play when the show wasn't on the air.

This live-action series was a surprisingly good show, with very good CGI (computer-generated imagery) special effects. Some

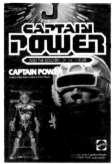

CAPTAIN POWER

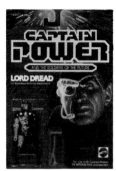

LORD DREAD

very important science-fiction writers worked on it, including J. Michael Stracynski, creator of the television show **Babylon Five**, and Larry DiTillio, also of **Babylon Five**.

The series was set in 2147, after the Metal Wars, which the machines had won. Lord Dread, looking very much like a Borg, was a human who had merged with a machine. He was Captain Power's arch-enemy, and "evil mas-

termind of the Bio Dread Empire." Before the war, he and Captain Power had been best friends. Captain Power and his group continued to fight the machines largely on their own. They were able to continue the battle because they had special armor that magically appeared out of thin air when they "powered up."

Both figures are fully articulated and come with weapons and the special armor.

CHARLIE'S ANGELS

HASBRO, 1977
SIZE: 8½"
VALUE RANGE: $10–$48

JILL
SABRINA
KELLY
KRIS

In the late 1970s, feminism had a toothy smile and big hair, and its name was Farrah. Even though it is certainly part of the tradition of "jiggle TV," Aaron Spelling's **Charlie's Angels**, arguably, was a landmark TV show. The premise is simple. Three smart and beautiful detectives come out of the police academy near the top of their class but then are assigned menial jobs by the patriarchy. Charlie Townsend sees their intrinsic worth and brings them to his

JILL

SABRINA

KELLY

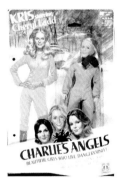

KRIS

detective agency so these "three little girls" could do some real police work.

The show made Farrah Fawcett-Majors a household name, and Jaclyn Smith, Kate Jackson and Cheryl

Ladd all stars. These 8½" action figures are really dolls, but we include them here because they portray the characters, not the actresses. Jill, Sabrina, Kelly and Kris all come on blister packs with a wrap-around logo, or in boxed gift sets. When Farrah Fawcett-Majors left the show, and Cheryl Ladd joined the cast, the logo changed. The original dolls all wore jumpsuits and boots. Many fashion accessories and "escapades sets" were available.

CHIPS

MEGO, 1978–81; LJN, 1983
SIZE: 8"
VALUE RANGE: $10–$45

JON
PONCH
SARGE

These action figures were based on the hit television show **CHiPs**, which starred Erik Estrada and told the adventures of three California Highway Patrol officers. The 8" Mego figures came with nicely detailed cloth uniforms and accessories like a pistol, holster and gun

JON

belt, and various decals. They are bubble-packed on cards. The packages show a color photograph of Jon and Ponch from the show.

PONCH

SARGE

CLASH OF THE TITANS

MATTEL, 1980
SIZE: 4"

PERSEUS
CALIBOS
THALLO
CHARON
PEGASUS (6¼")
KRAKEN (14⅜")

These imaginative, hard-to-find action figures are based on the 1980 MGM movie, **Clash of the Titans**. An "epic motion picture extravaganza," this movie featured the artistry of special effects master Ray Harryhausen. Harryhausen created the creatures and all effects for this movie without computer-generated imagery (CGI). He used models and miniatures.

The movie was based on tales from Greek

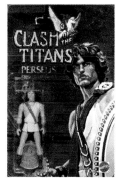

PERSEUS/$10–$50

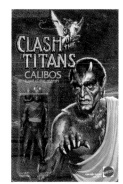

CALIBOS/$10–$50

THALLO/$10–$50

CHARON/$10–$50

mythology. Harry Hamlin played Perseus, the hero son of Zeus. Perseus, astride his winged horse Pegasus, aided

by his ally, Thallo, captain of the guard, battled Calibos, the evil lord of the marsh, Charon, the "Devil's

PEGASUS/$15-$45

boatman" and Kraken, a huge sea monster.

These action figures are nicely detailed. The faces convincingly convey the evil of Charon and Calibos, and the heroics of Perseus and Thallo. The human-shaped action figures in this series are all 4" and fully poseable. They all include accessories. Perseus and Thallo have swords and shields, and Charon and Calibos have swords. They are

KRAKEN/$65-$165

blister-packed on cards with a color illustration of the movie character. Pegasus is 6¼" high and has two removable wings. He is packaged in a window box. Kraken, the sea monster, is 14⅜" tall and approximately 20" long. His arms can be

posed, and he rotates at the waist. Kraken's tail can swing from side to side as well. Both tail and fins are detachable. Kraken came in a box with a full-color illustration of the monster from the film.

COMIC ACTION HEROES

MEGO, 1975–78
SIZE: 3¾"
VALUE RANGE: $12–$95

AQUAMAN
WONDER WOMAN
SHAZAM!
SUPERMAN
BATMAN
ROBIN
THE JOKER
GREEN GOBLIN
SPIDER-MAN
CAPTAIN AMERICA

The very first series of 3¾" action figures incorporates Marvel and DC comic-book heroes. Although they are nothing spectacular in terms of craft or design, these figures are immensely valuable to collectors. The figures all have bent knees, which are not articulated. The heads turn, and the arms can be raised and lowered. All are bubble-packed, and some have color pictures of other characters in the series on the front of the cards.

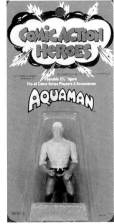
AQUAMAN

WONDER WOMAN

SHAZAM!

SUPERMAN

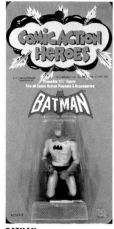

BATMAN

ROBIN

THE JOKER

GREEN GOBLIN

SPIDER-MAN

CAPTAIN AMERICA

CONEHEADS

PLAYMATES, 1993
SIZE: 6"
VALUE RANGE: $3–$10

BELDAR

Coneheads was a movie that tried to make the terms "consume mass quantities" and "we're from France" into pop cultural slang, but unfortunately, it didn't succeed. It was based on a series of hilarious, off-the-wall skits from the original **Saturday Night Live**. With Dan Aykroyd as Beldar, the male "parental unit," and most of the original cast in the movie, it was surprising that the film was not a hit. Most critics felt that the skits simply did not translate into a feature film, and the audience agreed. However, **Coneheads** does have its fans and a small cult-following.

The Beldar figure bears a close resemblance to Dan Aykroyd. The "reformed polymer replicant" is fully articulated and shares the grimace the character often wore in the movie. He is dressed in full-flight uniform

BELDAR

and comes with accessories. He is bubble-packed with a card photo of Jane Curtin, as Prymaat, and Beldar, urging everyone to "consume mass quantities."

COPS (N'CROOKS)

HASBRO, 1988
SIZE: 7"
VALUE RANGE: $5–$20

A.P.E.S.
SGT. MACE

These cyber-punk styled figures are cops who "fight crime in a future time," and it looks like that future includes "RoboCop"– type law enforcement officers. A.P.E.S. (Automated Police Enforcement Systems Officer) comes with a cap-firing rifle and power grabber. Sgt. Mace is a S.W.A.T.

Leader. He comes with a very large cap-firing bazooka. Both are fully articulated and detailed to look very menacing. The packaging shows stylized cybercops carrying huge weapons.

A.P.E.S.

SGT. MACE

DAKIN AND DAKIN STYLE FIGURES

R. DAKIN, 1965-77
SIZE: 8"
VALUE RANGE: $15–$40

BIG BOY
SMOKEY BEAR
BUGS BUNNY
WILE E. COYOTE
ROAD RUNNER
TWEETY BIRD
SYLVESTER
BARNEY RUBBLE
FRED FLINTSTONE
PEBBLES
BAMM-BAMM
YOGI BEAR
SCOOBY DO
MIGHTY MOUSE
MICKEY MOUSE
MINNIE MOUSE

GOOFY
DONALD DUCK
POPEYE
OLIVE OYL

R. DAKIN, 1965–77;
I.A.SUTTON, 1970-72
SIZE: 8"
VALUE RANGE: $100–$150

I.A. SUTTON
BANANA SPLITS:
  FLEAGLE BEAGLE THE DOG
  SNORKY THE ELEPHANT
  DROOPER THE LION
  BINGO THE BEAR

The plush fuzzy animals, promotional figures, and comic characters, of Dakin and Co. of San Francisco, are favorites of

BIG BOY

collectors worldwide. The advertising and promotional figures series, started in 1965, continued for a little over ten years and in-

SMOKEY BEAR

BUGS BUNNY

WILE E. COYOTE

ROAD RUNNER

cluded such beloved icons as the Big Boy, Smokey the Bear and Woodsy Owl. These figures were hollow plastic and ranged from 6″ to 9″. The Big Boy was all plastic, but Smokey and Woodsy had cloth outfits.

TWEETY BIRD

SYLVESTER

Dakin's T.V. Cartoon Theater series included favorite characters from Looney Tunes and Rocky and Bullwinkle. Bugs, Daffy and the rest of the "toons" personalities really shine through in these color-

ful figures. The Cartoon Theater characters came in colorful window boxes with the opening resembling a small television screen.

Dakin also produced a series of Hanna-

Barbera cartoon characters and Disney characters. Mickey Mouse, Minnie Mouse, Donald Duck, Goofy, The Flintstones, Yogi Bear, and Scooby Doo were among these. Popeye, Olive Oyl

BARNEY RUBBLE

FRED FLINTSTONE

PEBBLES

BAMM-BAMM

MICKEY MOUSE

MINNIE MOUSE

GOOFY

DONALD DUCK

YOGI BEAR

and Mighty Mouse also were part of the Dakin figure family. As with the Looney Tunes characters, these figures wonderfully convey the fun and whimsy of the TV cartoon characters.

Some figures had cloth costumes or other attachments. They were sold in re-closable bags with plastic handles. Some were also mounted on a variety of plastic bases as award-style

SCOOBY DOO

MIGHTY MOUSE

POPEYE

OLIVE OYL

DROOPER THE LION

BINGO THE BEAR

FLEAGLE BEAGLE THE DOG

SNORKY THE ELEPHANT

greetings called "Goofy Grams."

Dakin's craftspeople and designers were masters at re-creating the special appeal of a character's personality. The superior sculpting did credit to Dakin, which was one of the few companies to produce Warner Brothers, Disney and Hanna-Barbera characters. As the line changed over the years, selected Disney characters were produced in two sizes.

Most advertising figures were one-shot promotions, some only in certain regions, and are therefore rarer.

Dakin's success with these benign, popular toys spawned a number of competitors, most notably I.A. Sutton. Sutton successfully mimicked the Dakin style with their Banana Splits figures. These characters from the popular 1970 Hanna-Barbera live-action children's TV show—Fleagle, Bingo, Drooper and Snorky—were favorites among today's Gen-Xers and younger baby boomers. Original Banana Splits figures are highly desirable.

The tremendous success of **Star Wars** probably helped to end this popular figure style, but any of the nostalgic figures make an attractive addition to toy collections.

DARK KNIGHT COLLECTION

KENNER, 1990–91
SIZE: 5"

**CRIME ATTACK BATMAN
BRUCE WAYNE
KNOCK-OUT JOKER**

The tremendous popularity of the 1989 **Batman** feature film created a new demand for Batman action figures. Since the **Batman** name and logo were already in use on figures being produced by Toy Biz, Kenner developed the Dark Knight Collection. Although it was not a direct movie tie-in, it was, nevertheless,

CRIME ATTACK BATMAN/ $5–$20

BRUCE WAYNE/$5–$18

very popular. All the figures were carefully sculpted, high-quality toys and far superior to the Toy Biz line. It gave Kenner a foot in the door with Warner Bros., and Kenner

eventually won the license for **Batman Returns** and other major WB productions.

In DC Comics, **The Dark Knight** stories

KNOCK-OUT JOKER/$30–$100

were credited with giving Batman his darker image. They told the tales of Bat-man's earlier cases and are now considered part of the "Else-worlds" stories, that is, stories that may or may not have happened to Batman. But in this case, these Batman figures are related to the movie.

Crime Attack Batman comes with a Bat-a-rang and claw and is dressed in the batsuit and removable cape. Bruce Wayne is dressed in a black workout suit and comes with a "quick change suit," which enables him to turn into Batman. He also comes with weapons, a clicker communicator, and a bat belt. Knock-Out Joker is dressed in his usual bright colors and comes with an over-sized bazooka and an accessory pistol that shoots the stylized "POW." All three figures are bubble-packed. The cards have color pictures of the characters.

DARKWING DUCK

PLAYMATES, 1991
SIZE: 5"
VALUE RANGE: $2–$8

**DARKWING DUCK
LAUNCHPAD McQUACK**

Disney's Darkwing Duck is a crime-fighting duck from St. Canard whose slogan is "I am the terror that flaps in the night!" This serio-comic super hero duck has a side-kick, Launchpad Mc-Quack, who's none too bright, but is great with machines. Together, the feathered heroes fight criminals with names like Tuskerninni.

DARKWING DUCK

LAUNCHPAD McQUACK

This animated TV series was great fun in the best Disney tradition, and Playmates captured the spirit of the show in the action figures. Both came with unique action features. Darkwing, with a spinning hat and cape; Launchpad with a spinning head. Both come with ray guns. They are bubble-packed on cards with color pictures of Darkwing Duck as he appeared in the cartoon.

DC COMICS SUPER HEROES

TOY BIZ, 1989–90
SIZE: 5"
VALUE RANGE: $10–$35

SUPERMAN WITH KRYPTONITE RING

WONDER WOMAN

THE PENGUIN, MISSILE-FIRING

THE RIDDLER, WITH RIDDLES AND CLUES

ROBIN, KARATE CHOP

SUPERMAN WITH KRYPTONITE RING

WONDER WOMAN

The enduring heroes of DC Comics appear again as action figures in this colorful series. All are fully poseable and come bubble-packed on cardboard with a color picture of the character as she or he appears in the DC Comic.

Each figure comes with a unique action feature. Missile-firing Penguin originally fired a "safe soft missile," but it didn't meet federal safety standards. Later Penguin figures fired larger missiles or an umbrella that

continued on page 85

The Super Deal

DC Comics asked the Kenner presentation team to represent DC's super hero action figure license. Little did the Kenner team know what a valuable opportunity it had been afforded. Kenner was the obvious choice for DC for two reasons: Mego, Kenner's former license went out of business in 1982; and Kenner's **Star Wars** figures were writing a new chapter in action figure history. The industry was also heating up as Kenner went head on against Mattel who was riding high with its **Masters of the Universe** line.

Kenner pitched its line as "DC Super Heroes with the Super Power Feature." They hadn't worked any of these features out in detail, but got the license for what ended up to be the **Super Powers** line.

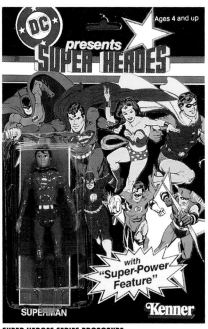

SUPER HEROES SERIES PROTOTYPE

Creating its relationship with DC Comics turned out to be immensely critical for Kenner. In 1980, the success of the **Batman** film went high above and beyond all expectations. Kenner did not land this license right off, but when Toy Biz couldn't keep up with demand, Kenner was DC's obvious choice. **Batman** figures were based on the **Dark Knight** comic books and for over twenty years Kenner has produced nearly 150 different figures. The right to the enormous license for continuous production of **Batman** figures was a result of Kenner's cleverness, highly successful track record and great timing.

THE PENGUIN, MISSILE-FIRING

THE RIDDLER, WITH RIDDLES AND CLUES

ROBIN, KARATE CHOP

"pops its top" when activated.

Wonder Woman's "power arm" throws the "Golden Lasso of Truth." The Riddler is equipped with six preprinted riddles that

are clues to his evil deeds. Robin has a button-activated karate chop and comes complete with a Batarang, grapple line and detachable cape. The coolest of

the bunch may be Superman. He has "magnetic action" that causes him to falter when exposed to Kryptonite, and he comes with a kid-sized Kryptonite ring.

DEFENDERS OF THE EARTH

GALOOB, 1986
SIZE: 6"
VALUE RANGE: $5–$25

FLASH GORDON
THE PHANTOM
MANDRAKE THE MAGICIAN
MING THE MERCILESS

These figures were based on a 1986 King Features Syndicate animated series. It brought together an unusual combination of characters from the early days of King Features action adventure. The premise is that Flash Gordon, the

FLASH GORDON

THE PHANTOM

MANDRAKE THE MAGICIAN

MING THE MERCILESS

Phantom, and Mandrake the Magician teamed up in the year 2015 to keep Ming the Merciless from taking over the world. They got some help from Gordon (Flash's son), Jedda Walker (the

Phantom's daughter), L.J. (Lothar's son), Kshin (an orphan), and Zuffy (a furry alien).

Each colorful figure is fully articulated and has a knob on its back

that activates the "power punch." The brightly colored card has a picture of the three heroes as they appear in the cartoon—and a larger picture of the specific figure.

DIE-CAST SUPER HEROES

MEGO, 1979
SIZE: 5½"
VALUE RANGE: $20–$150

SPIDER-MAN
BATMAN
HULK
SUPERMAN

In the 1970s Mego dominated the action figure industry. Its success was made on its super heroes, TV and film-related toys, and **Micronauts**. In 1979, the 30th anniversary

of the company, Mego decided to expand heavily into other toy areas. But it was the wrong year to neglect action figures. Kenner's **Star Wars** action figures quickly took over the market, and

SPIDER-MAN

BATMAN

HULK

Mego was out of the toy business before the end of the year.

The **Die-Cast Super Heroes** was one of the ideas Mego developed while in its death-throes. These high-quality toys were limited-edition, die-cast metal and plastic figures. They were really limited to the number Mego could sell for one-time distribution. It is unknown how many of each fig-

SUPERMAN

ure were produced, but they are difficult to find. The heads are plastic, and the costumes are cloth. The metal bodies are

nicely detailed and brightly painted, and all figures are fully articulated. These 5½" figures are packaged in a combination window box/card. There is a color picture of the specific Marvel character on the box, and the logo of the super hero.

DICK TRACY COPPERS AND GANGSTERS

PLAYMATES, 1990
SIZE: 5"

**DICK TRACY
THE TRAMP
THE BLANK**

DICK TRACY/$5-$20

THE TRAMP/$5-$20

This series appeared shortly before the 1990 film **Dick Tracy**. Directed by and starring Warren Beatty, and based on the Chester Gould strip, the film is more like a comic book than any of the comic-book films of the late 1980s and 1990s, with a truly outstanding visual design that utilizes an elaborate and sophisticated use of matte paintings and color. But audiences were expecting a **Batman**-like epic, and were turned off by the relatively low-key and old-fashioned approach to crime fighting found in **Dick Tracy**. Even Madonna

as Breathless Mahoney and cameos by Al Pacino and Dustin Hoffman couldn't make this a successful romp commercially. But the visuals of the film are still quite stunning.

The action figures are closer to the comic-strip characters than they are to the film's. Dick Tracy, the Tramp and even the Blank look as if they stepped off the comic page rather than the silver screen. Dick Tracy comes wearing his trademark fedora, and he's dressed in suit vest and shirt-

sleeves. With his billy club, pistol and rifle, he's ready for anything. The Tramp, one of the robbers who lurks in the city sewers, is dressed in suitable ragtag clothes, with a porkpie hat. His weapons include a nasty-looking shiv, a big plank of wood, and a manhole cover.

An organization for the homeless tried to get The Tramp recalled. Whether it was positive or negative, this conflict created much publicity. Dealers anticipated the demand, scooped up

THE BLANK/$50–$180

every available figure in this series and commanded extra money for this "unbelievable" figure. Playmates Toys also benefitted greatly as the commotion reignited interest in the line for several more weeks.

The Blank, perhaps the most interesting of the bunch, comes with a removable blank face mask, a pistol, and a briefcase. This figure was sold only in Canada and was initially withheld because when the mask is removed, it reveals Madonna, thus giving away the surprise ending of the movie. By the time the figure was cleared for release, the movie had already bombed and **Dick Tracy** merchandise was dead. The Blank probably would never have been released were it not for a Sears Canada contract that had to be fulfilled up to Christmas 1990. Action-figure dealers bought up as many as they could and filtered them south to the United States.

DOCTOR WHO

DENYS FISHER TOYS, 1976
SIZE: 8"
VALUE RANGE: $100–$225

DOCTOR WHO
LEELA

The longest-running science-fiction series on television is not **Star Trek**, but the British import **Doctor Who**. The BBC first aired it as a children's show in 1963, and it has been on the air in the UK and the United States in one form or another since then. Most recently, **Dr. Who** was a made-for-television movie in 1996 and a new radio drama in the United Kingdom.

The show chronicles the adventures of Doctor Who, an alien Time Lord from the planet Gallifrey, who, in direct contradiction of his people's laws, steals a time machine —a TARDIS—which is now stuck in the shape of a blue British police box because of a malfunction. The

DOCTOR WHO

Doctor begins exploring the universe, often in the company of friends he meets

LEELA

along the way. He aids the oppressed, fights injustice and is generally an all-around de-

cent guy. He is joined by his beautiful companion, Leela; K-9, his robotic dog ("Dr. Who's best friend"); and other aliens he meets in his adventures in time and space.

The Dr. Who action figures were manufactured by Mego and marketed in the UK by Denys Fisher Toys.

DUKES OF HAZZARD

MEGO, 1981
SIZE: 8"
VALUE RANGE: $10–$50

BO
LUKE
DAISY

The Dukes of Hazzard was a summer trial replacement show that aired on CBS in 1979. With the pilot episode "One Arm Bandits," the

BO

show was an instant success. It stayed on the air until 1985. The show followed the exploits of Hazzard

LUKE

County residents, cousins Bo and Luke, and their car, the General Lee. They were always up against a

DAISY

typical redneck sheriff, Boss Hogg, and his dim-witted deputy. The boys have a helpful uncle, and a beautiful cousin Daisy, who has her own Jeep.

The action figures were made in two sizes—8" and 3½".

The larger figures are costumed in cloth, and the smaller are all molded plastic. Both sizes are fully articulated and bubble-packed on a card with a color picture of the main characters from the show.

DUNE

LJN
SIZE: 6"
VALUE RANGE: $10–$30

PAUL ATREIDES
FEYD
SARDAUKAR WARRIOR

In 1984, a big-budget film based on Frank Herbert's science-fiction epic **Dune** hit the big screen. It was directed by offbeat, cult director David Lynch, cost a fortune, starred many unknown and unproven actors, was painstakingly produced, and

PAUL ATREIDES

FEYD

failed big at the box office.

Anyone who has seen the film may find it hard to imagine a child sitting through it, let alone asking for an

action figure based on any of the characters. But most merchandising deals are made long before a film reaches the theaters. The Dune figures lan-

SARDAUKAR WARRIOR

guished on toy store shelves.

The action figures included Paul Atreides, Feyd, and a Sardaukar Warrior. In the movie, Sting played Feyd, and this figure is sometimes priced higher than the others. One figure that was never produced was Gurney Halleck. One can only imagine what it would be worth today. That character was played by Patrick Stewart, better known as Captain Picard from **Star Trek: The Next Generation**.

E.T.

LJN, 1982–83
SIZE: 4½"
VALUE RANGE: $5–$16

E.T.

No one expected a movie about a crinkly alien and a little boy to be much of a success, so when there was a rush for E.T. merchandise, the figures weren't there. Steven Spielberg had demanded absolute secrecy during the making of the film, so no prototypes were made. The manufacturers couldn't produce the figure until the film was released, and the stores didn't want to order the action figure without having seen the figure. The secrecy backfired, and bootleg E.T.'s were popping up everywhere. By the time the authorized figures reached the stores, the interest had waned, and E.T. ended up on sale shelves, sometimes for years. Nevertheless, the figures are cute. There is a talking

E.T.

E.T., a wind-up E.T. that can "walk and waddle," and a fully poseable E.T. with an extending neck.

EVEL KNIEVEL

IDEAL, 1973–74
SIZE: 8"
VALUE RANGE: $10–$40

EVEL KNIEVEL

The death-defying daredevil inspired a fun line of action toys in the early 1970s. Although technically "bendies," these fig-ures were sold as action figures. Evel comes in a bubble-pack on a card, with a removable helmet and a "swagger stick." Some of the stunt vehicles also came with action figures.

EVEL KNIEVEL

FANTASTIC FOUR

TOY BIZ, 1994–95
SIZE: 5"
VALUE RANGE: $3–$10

MR. FANTASTIC
THE THING
INVISIBLE WOMAN
HUMAN TORCH
SILVER SURFER
NAMOR, THE SUB-MARINER
DR. DOOM

The animated television show, **Marvel Action Hour,** featured the Fantastic Four and other Marvel comic-book heroes. This series, based on the cartoon, marked the first time the Fantastic Four had an action figure series of their own. Previously, they had appeared in other Marvel comic-book hero toy lines. Toy Biz originally included only Mr. Fantastic and the Thing, but retailers demanded all four. Since the molds for the Human Torch and the Invisible Woman weren't ready yet, Toy Biz repainted the Sil-

MR. FANTASTIC

ver Surfer to create the Human Torch, and the Iron Man and Spider-Woman for

THE THING

INVISIBLE WOMAN

HUMAN TORCH

the Invisible Woman. These two figures were released in limited quantities until the new versions were ready.

The original Fantastic Four gained their superpowers while on a mission to deep space. Their rocket ship was struck with heavy cosmic rays. Scientist Reed Richards, his fiancée Sue Storm, her brother Johnny Storm,

and Reed's friend Ben Grimm absorbed massive radiation, which mutated them into beings with incredible powers. Reed Richards gained the ability to stretch his entire body into bizarre forms. Sue became able to turn invisible at will. Johnny was transformed into what "if . . .?" a flaming human torch. And Ben was transformed into a hulking, rock-

covered beast. He is the only member of the Four who cannot transform back to his human form. They returned to Earth as the Fantastic Four.

The Mr. Fantastic action figure has "super stretch arms." He is painted blue with white and brown hair. He is the leader of the four. The Thing comes with "clobberin' time punch," and his yellow body is muscle-

SILVER SURFER

NAMOR, THE SUB-MARINER

DR. DOOM

bound. The Human Torch figure is bright red with fluorescent flames painted all over his body. These are his "glow-in-the-dark flames." He also comes with a flaming shield. The Invisible Woman is dressed in a blue uniform similar to Mr. Fantastic's, and she comes with an invisible force shield made of clear plastic. Another version of the

Invisible Woman was manufactured in clear plastic.

Joining the Fantastic Four are more Marvel characters, including the Silver Surfer; Namor, the Sub-mariner; and bad guy Dr. Doom. The Silver Surfer is painted completely silver, comes with his space surfboard, and has "space surfing action." Namor is packaged with an

undersea trident and a shield. He has an action power punch. Dr. Doom is a sinister figure, with a hood over his skeletonlike face. He, too, has action power punch. All the figures are bubble-packed with the Fantastic Four logo on the card.

FLASH GORDON

MEGO, 1976
SIZE: 12"
VALUE RANGE: $40–$150

FLASH GORDON
DALE ARDEN
DR. ZARKOV
MING, THE MERCILESS

MATTEL, 1979
SIZE: 3¾"
VALUE RANGE: $8–$60

FLASH GORDON
DR. ZARKOV
THUN, THE LION MAN
MING, THE MERCILESS
BEASTMAN

FLASH GORDON

DALE ARDEN

DR. ZARKOV

MING, THE MERCILESS

Although Flash Gordon has been around since his first appearance in a comic strip in 1934, the 1970s saw an update of the character and his adventures. An animated TV series and a live-action movie brought Flash; Dale; Dr. Zarkov; Ming, the Merciless; and the Planet Mongo back into the cultural Zeitgeist. The Flash Gordon Mego line is based on the film; the Mattel line on the television series.

The four 12" Mego figures are finely detailed with cloth costumes. Flash wears a bright red shirt and a matching helmet; Dale wears a light yellow minidress with a red scarf. She also comes with a golden crown. Both Dr.

FLASH GORDON

DR. ZARKOV

THUN, THE LION MAN

MING, THE MERCILESS

BEASTMAN

Zarkov and Ming, the Merciless, have beards, but we know Ming is up to no good by his markedly sinister expression.

Dr. Zarkov has the same heroic pose as Flash and wears a similar helmet. All figures are bubble-packed on cards with a futuristic picture.

The 3¾" Mattel line offers the animated characters, many of which were exclusive to the Saturday morning TV series. The smaller Mattel figures are all plastic and are fully articulated. Flash is once again dressed in his red uniform and sports blond hair. Dr. Zarkov is somewhat barrel-chested, and with his whiskers and blue uniform looks more than a little like Mr. Spock of **Star Trek**. Thun, the Lion Man, has a lion's head on a muscular male body, and Beastman is a fierce-looking blue creature. In this series, Ming, the Merciless, is simply called "Ming." That he is pure evil is still evident in his expression, although he is not as finely detailed as the larger Ming. He does, however, more closely resemble the half man/half lizard creature that, in later versions of Flash Gordon, he would become. All figures are bubble-packed on a card illustrated with a scene from the cartoon.

THE FLINTSTONE KIDS

COLECO, 1987
SIZE: 3½"
VALUE RANGE: $8–$35

CAVEY, JR.
PHILO QUARTZ
WILMA SLAGHOOPLE

This Flintstones series is based on the Hanna-Barbera animated television show from 1986 called **The Flintstone Kids.** This spin-off of the original Flintstones follows the adventures of the original series' characters as children.

The action figures are as cute as the characters. Each figure is a miniature version of

CAVEY, JR.

PHILO QUARTZ

his or her "adult" Flintstone counterpart. All are dressed in the trademark Flintstone animal-skin outfits, although Cavey, Jr., is covered with long hair, and his clothing isn't visible. All the figures come with a Stone Age vehicle: Cavey, Jr., with his turtle shell rider; Philo

WILMA SLAGHOOPLE

Quartz with his Stone Age police car; and Wilma Slaghoople

with her prehistoric mastodon rider. Some of the figures are bubble-packed with their vehicle on a card illustrated with a picture of the specific character from the series; however, the cards for Wilma, Betty and Dino have the Flintstone Kids logo, but no pictures of the characters.

GARGOYLES

KENNER, 1995–96
SIZE: 6½"
VALUE RANGE: $5–$12

QUICK STRIKE GOLIATH
POWER WING GOLIATH
ELISA MAZA

This very fine series of action figures from Kenner was based upon the popular Disney animated TV series **Gargoyles**.

It is the story of New York City stone Gargoyles named Goliath, Brooklyn, Broadway, Lexington, Bronx and Hudson who come to life during the night and turn back into stone each day. The Gargoyles' true nature is known only to a small group of hu-

QUICK STRIKE GOLIATH

mans, including their friend and protector, Detective Eliza Maza, and the billionaire David Xanatos. In 1994, David Xanatos purchased the castle "lock, stock and gargoyle," moving it to the top of his skyscraper in Manhattan, not knowing the history of the castle.

POWER WING GOLIATH

In Scotland, in A.D. 994 the Magus of Castle Wyvern had cast a spell on the gargoyles, turning them to stone. Once living defenders of the castle, they slept in their stone form for a thousand years. By moving the castle "above the clouds," the thousand-year spell was broken, and the Gargoyles came back to life.

The Gargoyle action figures are nicely sculpted with details like oversize, clawed hands and varying facial expressions and fangs. All come with action features. Quick Strike Goliath is stone-gray and has a battle ax and "leaping attack action." Broadway, a hulking creature, is also stone-gray and

ELISA MAZA

has large batlike wings and a medieval mace. Brooklyn is

brownstone-colored and leaner than the other Gargoyles. He has a large, oversize sword. Power Wing Goliath, from the second series of figures, has mechanized wings and a sword. Elisa Maza, like any good NYPD detective, comes with several large weapons and a blue suit with wings so she can share the skies with the Gargoyles.

GHOST RIDER

FLEETWOOD TOY COMPANY, 1976
SIZE: 100"
VALUE RANGE: $40–$300

GHOST RIDER, MOLDED ONE-PIECE TOY

Ghost Rider first appeared in a Marvel comic in the 1970s. The Fleetwood action figure was a one-shot toy based on this character. He comes with a motorcycle with "super fast wheels" and two faces—one of his Ghost Rider identity, and one of his human identity, Dan Ketch. Ghost Rider was revived in the 1990s, as a comic-book hero and, by Toy Biz, as an action figure.

GHOST RIDER

12" GI JOE

HASBRO, 1964–78
SIZE: 12"

ACTION SOLDIER
ACTION SAILOR
ACTION MARINE
ACTION PILOT
WEST POINT CADET
PHOTO BOX
MILITARY POLICE
PHOTO BOX
SKI PATROL
SPECIAL FORCES
GREEN BERET
ACTION SOLDIER
(AFRICAN-AMERICAN)
TALKING ACTION
SOLDIER
SEA RESCUE
FROGMAN
SHORE PATROL SET
DEEP SEA DIVER
LANDING SIGNAL
OFFICER
ANNAPOLIS CADET
TALKING ACTION SAILOR
COMMUNICATIONS POST
& PONCHO SET
COMBAT PARATROOPERS
MARINE MEDIC SET
DRESS PARADE SET
AIR CADET IN PHOTO
BOX
FIGHTER PILOT IN
PHOTO BOX
TANK COMMANDER
JUNGLE FIGHTER
TALKING ACTION MARINE
ASTRONAUT SUIT

TALKING ACTION PILOT
GERMAN STORM
TROOPER
JAPANESE IMPERIAL
SOLDIER
RUSSIAN INFANTRY MAN
FRENCH RESISTANCE
FIGHTER
BRITISH COMMANDO
AUSTRALIAN JUNGLE
FIGHTER
ACTION NURSE WITH
WHITE BAG
TALKING GI JOE
ASTRONAUT
TALKING ADVENTURE
TEAM COMMANDER
CANADIAN MOUNTIE
GIFT SET
KUNG FU GRIP LAND
ADVENTURER
BLACK ADVENTURER
MOVING EYES LAND
COMMANDER
ATOMIC MAN
BULLETMAN
DANGEROUS REMOVAL
SMOKE JUMPER
KARATE
EMERGENCY RESCUE
SECRET AGENT
SUPER JOE

ACTION SOLDIER/$100–$300

GI Joe was released in 1964 by Hasbro as an attempt to create a fighting figure for the boy's toy market—figures that were never, ever to be referred to as "dolls." The original GI Joe figures were fully poseable, and 12" tall and represented each of the four branches of the armed services. Action Soldier, Action Sailor, Action Marine and Action Pilot were the first to hit the shelves and sold out entirely. Accessory sets—including tents, mess kits, weapons, sleeping bags, life rafts, parachutes and many other kinds of military gear—were created for each figure and sold separately. Each figure came with

ACTION SAILOR/$75–$350

ACTION MARINE/$80–$335

ACTION PILOT/$125–$400

**WEST POINT CADET PHOTO BOX/
$225–$1,100**

**MILITARY POLICE PHOTO BOX/
$650–$1,900**

a scar on the right cheek, one of four standard hair colors, the uniform appropriate to the branch of the armed forces it represented, accessories and boots. The first-issue boots are distinguishable from

later issues because they are made of rubber and had smooth bottoms, instead of the ridged bottoms of the later-issue boots. Also, the feet of the GI Joe figures in the first issue were manufactured in a smaller

size than the feet of later figures.

All of the original figures are extremely valuable to collectors. Action Pilot is the most treasured, followed by Action Sailor, Action Marine,

SKI PATROL/$155–$850

SPECIAL FORCES/$135–$800

GREEN BERET/$200–$400

ACTION SOLDIER (AFRICAN-AMERICAN)/$500–$1,400

TALKING ACTION SOLDIER/ $150–$375

SEA RESCUE/$75–$250

and Action Soldier. Most of the equipment or accessory kits are also sought after, and certain sets are worth thousands of dollars.

Hasbro modified the placement of the GI Joe logo, ™, and ® symbols over the course of production, resulting in twelve different boxes—three different styles of boxes for each of the four original figures. The boxes included different field manuals for each of the four figures and were color-coded based upon the figure. Action Soldier came with a light beige manual, Action Marine with an olive green manual, Action Sailor with a dark blue one,

FROGMAN/$135–$500

SHORE PATROL SET/ $120–$450

DEEP SEA DIVER/$150–$1,200

LANDING SIGNAL OFFICER/ $125–$600

ANNAPOLIS CADET/$250–$900

TALKING ACTION SAILOR/ $160–$750

and Action Pilot with a light blue one. (The manuals that came in the narrow window boxes were smaller than the manuals that came with the accessory cards.) The manuals included information on posing the figure, dressing

the figure, removing the boots, and where to place the sticker insignia.

More specialized GI Joe figures followed, riding on the success of the original four figures. West Point Cadet, Military Police, Mountain Troops, Ski

Patrol, Special Forces, and Green Beret GI Joe outfits were all accessorized and detailed to represent further divisions within the U.S. Army— details that are reflected in their high values. The Mountain Troops GI Joe set fea-

COMMUNICATIONS POST & PONCHO SET/$175–$225

COMBAT PARATROOPERS/ $40–$160

MARINE MEDIC SET/$90–$400

DRESS PARADE SET/$75–$425

AIR CADET IN PHOTO BOX/$100–$400

tured white snowshoes, a backpack, grenades and a rope, while the Ski Patrol set came with a white snowsuit, skis, ski boots and white mittens.

An African-American Action Soldier was produced, and a talking Action Soldier that "spoke" was introduced in 1967. When Joe's dog tag was pulled it said such things as "Cover me," "Commence firing" and "Send reinforcements."

Sea Rescue, Frogman, Shore Patrol and Deep Sea Diver equipment could be purchased for Action Sailor. The sets included highly detailed and unique accessories and outfits, like the Deep Sea Diver in a diving suit with helmet, and Frogman's scuba suit complete with tank, fins

FIGHTER PILOT IN PHOTO BOX/
$400–$1,900

TANK COMMANDER/
$150–$1,700

JUNGLE FIGHTER/$225–$2,000

TALKING ACTION MARINE/
$60–$875

ASTRONAUT SUIT/$80–$1,000

and depth gauge. An Annapolis Cadet set was also produced, with a full-dress uniform for Action Sailor. A Talking Action Sailor similar to the Talking Action Soldier was manufactured, as well.

Action Marine's Communications set included a field radio, glasses, telephone, maps and map cases. The Paratroopers set was also highly specialized with extraordinary attention to detail in the weapons and equipment. Marine Dress was simple, but the loaded Medic set certainly was sought after by children, who reveled in dressing their Action Marines in the Medic's medical accessories and white armbands with red crosses. Tank Commander and Jungle Fighter are also notable sets—Tank Commander's "leather"

TALKING ACTION PILOT/
$200–$850

GERMAN STORM TROOPER/
$400–$1,700

JAPANESE IMPERIAL SOLDIER/
$500–$1,900

RUSSIAN INFANTRY MAN/
$350–$1,600

FRENCH RESISTANCE FIGHTER/
$280–$1,500

BRITISH COMMANDO/
$300–$1,600

jacket with fur trim serves as a perfect example of the attention to detail representative of the equipment sets. As with the other figures, a Talking Action Marine was produced

that could say "Dig in," "Move out," "Geronimo" and other short phrases.

A Talking Action Pilot that came in the trademark orange jumpsuit and "spoke" phrases like "All systems go"

was part of the Action Pilot collection. Like the others, Action Pilot had an Air Cadet accessory set with uniform. The extremely valuable Fighter Pilot came with a working parachute and is

AUSTRALIAN JUNGLE FIGHTER/ $225–$1,550

ACTION NURSE WITH WHITE BAG/$1,200–$3,000

TALKING GI JOE ASTRONAUT/ $140–$550

TALKING ADVENTURE TEAM COMMANDER/$90–$120

worth over $1,900 in mint condition in its original packaging. The timely Astronaut set comes with a space suit, oxygen mask and tether cord, while the Air Sea Rescue set included a lifesaver and an orange scuba suit.

In 1966, Hasbro released a series of Action Soldiers of the World and their accessories. The series included a German Storm Trooper, a Japanese Imperial Soldier, a Russian Infantry Man, a French Resistance Fighter, a British

Commando and an Australian Jungle Fighter.The scarred cheeks were gone, and the heads had been replaced—with the exception of the Japanese soldier— with thinner, more European-looking faces. The Japanese soldier had a different head with slanted eyes and always came with black hair, while the other figures came in all of the standard GI Joe hair colors. Sometimes the figures were packaged with original GI Joe scarred heads, but the unscarred figures are preferred by collectors. Also, there is a less common Japanese soldier whose body has a yellowish tinge—in all likelihood a test run of a color that was never widely released by Hasbro.

Each of the figures was dressed in a uniform specific to its nationality and came with multiple accessories. They were packaged with their accessories in display boxes but were also sold separately. Hasbro of Canada produced a Canadian Commando, along with the other six figures, which came in the same box as British Commando but was designated as "British and Canadian Commando" on the box.

The GI Joe Action Nurse sets featured a female doll dressed in a white nurse's outfit and came with crutches, bandages, splints, a stethoscope and a plasma bottle. "America's Moveable Action Girl" came in sets with white or green shoulder bags. The Nurse with the white shoulder bag is almost three times as valuable

CANADIAN MOUNTIE GIFT SET/ $300–$2,000

KUNG FU GRIP LAND ADVENTURER/$60–$140

BLACK ADVENTURER/ $100–$225

MOVING EYES LAND COMMANDER/$50–$150

as the Nurse with the green bag. The Mountie Gift set featured a Royal Canadian Mounted Police uni-

form and came with various accessories including snowshoes and a climbing stick.

The Adventures of GI Joe series included Talking GI Joe Astronaut, whose phrases spoke to America's pride in space exploration. Adventure Team GI Joe figures, Kung Fu Grip Land Adventurer and Adventure Team Commander advertised the character's "lifelike hair and beard." The muscular Black Adventurer was "lifelike," and wore only a pair of camouflage shorts.

A set of Eagle Eye GI Joes with moving eyes included Moving Eyes Land Commander, with the same hair and beard as the Adventure Series figures.

Atomic man figures had a metallic right arm and left leg. Billed as the "newest member of the adventure team," he wore a short-sleeved camouflage shirt and brown

ATOMIC MAN/$20-$65

BULLETMAN/$55-$175

DANGEROUS REMOVAL/ $30-$75

SMOKE JUMPER/$30-$90

pants and came packaged in a display box with a card advertising his "atomic flashing eye." Bulletman, "the human bullet," wore an orange suit and orange boots with a black belt and came packaged on a card painted like a stone wall.

Hasbro's creativity was never-ending—Dangerous Removal GI Joe sets were sold with armor suits, face shields, bombs and bomb detonation boxes. Smoke Jumper and Karate figures were included in the same series, along with Emergency Res-

KARATE/$40–$120

EMERGENCY RESCUE/$40–$90

SECRET AGENT/$40–$90

SUPER JOE/$5–$30

well. It wasn't the GI Joe people knew and loved, so the Super Joe figures were pulled from the market after two years.

Super Joes with "1-2 punches" were produced in both Caucasian and African-American figures. The Caucasian Super Joe wore a yellow jumpsuit, and the African-American Super Joe wore orange. They were packaged on updated cards featuring cartoon representations of the figures. Night Fighters Luminos was part of the Super Joe series, and the figure's clear body lit up when operated on a AA battery. The Shield carried a metallic shield, was also part of the Super Joe Adventure Team, and illustrates how far Hasbro had come since the inception of their "American fighting man."

cue and Secret Agent sets. Each set featured a fitting outfit, and came with "Authentic equipment for Adventure Team." Secret Agent even had a trench coat and a bulletproof vest.

The Arabian Oil crisis during the 1970s caused the price of plastic to rise sharply.

To combat escalating costs the size of GI Joe was reduced to 8 inches and named "Super Joe." These new Super Joe figures were not compatible with the previous 12-inch scale outfits. Other unsuccessful elements were joints that broke easily and a new space-type theme that did not sell

G.I. JOE— A REAL AMERICAN HERO

HASBRO, 1982–94
SIZE: 3¾"

INFANTRY TROOPER— GRUNT

RANGER—STALKER

COMMANDO—SNAKE EYES

MORTAR SOLDIER— SHORT FUZE

COMMUNICATIONS OFFICER—BREAKER

MACHINE GUNNER— ROCK 'N ROLL

BAZOOKA SOLDIER—ZAP

COUNTER INTELLIGENCE— SCARLETT

LASER RIFLE TROOPER— FLASH

COBRA OFFICER

COBRA COMMANDER

COBRA

MEDIC—DOC

S.E.A.L.—TORPEDO

MINE DETECTOR— TRIPWIRE

MARINE—GUNG-HO

ARCTIC TROOPER— SNOW JOB

HELICOPTER ASSAULT TROOPER—AIRBORNE

MAJOR BLUDD

DESTRO

FIRST SERGEANT—DUKE

HEAVY MACHINE GUNNER—ROADBLOCK

TRACKER—SPIRIT

DOG HANDLER—MUTT & JUNKYARD

COBRA INTELLIGENCE OFFICER—BARONESS

COBRA SABOTEUR— FIREFLY

COBRA NINJA—STORM SHADOW

FIRST SERGEANT—DUKE

ZARTAN

INFANTRY TROOPER— FOOTLOOSE

SILENT WEAPONS— QUICK KICK

COVERT OPERATIONS— LADY JAYNE

COMMANDO—SNAKE EYES II

SAILOR—SHIPWRECK

WARRANT OFFICER— FLINT

COBRA FROGMAN—EELS

COBRA POLAR ASSAULT— SNOW SERPENT

CRIMSON GUARD COMMANDERS— TOMAX & XAMOT

ICEBERG

SCI-FI

ZANDAR

ZARANA

LEATHERNECK

WET-SUIT

HAWK

B.A.T.

DR. MINDBENDER

COBRA COMMANDER

CRYSTAL BALL

BIG BOA

JINX

RAPTOR

CROC MASTER

GUNG-HO (DRESS UNIFORM)

BLOCKER

AVALANCHE

MAVERICK

DODGER

SHOCKWAVE

BUDO

VOLTAR

SPEARHEAD & MAX

STORM SHADOW II

HOODED COBRA COMMANDER

SGT. SLAUGHTER

THE FRIDGE

SERPENTOR

SNAKE EYES III

T.A.R.G.A.T.

GNAWGAHYDE

LIFELINE

RED STAR

DESTRO II

WILD BILL

EFFECTS

LOBOTOMAXX

PREDACON

CARCASS

30TH ANNIVERSARY ACTION SOLDIER

30TH ANNIVERSARY ACTION SAILOR

30TH ANNIVERSARY ACTION MARINE

30TH ANNIVERSARY ACTION PILOT

30TH ANNIVERSARY ORIGINAL ACTION TEAM

After a four-year hiatus from production, Hasbro came out with a series of 3¾" G.I. Joe figures that became the largest line of action figures ever produced. The new figures were influenced by the change in action figure size brought along after the release of the Star Wars line in 1978 and were much smaller than the 12" G.I. Joe figures of the past. G.I. Joes were designated "G.I. Joe—A Real American Hero" and were named not only by their classification, as the original Joes had been, but

INFANTRY TROOPER—GRUNT/ $20–$90

RANGER—STALKER/$30–$100

COMMANDO—SNAKE EYES/ $35–$200

MORTAR SOLDIER— SHORT FUZE/$20–$90

COMMUNICATIONS OFFICER— BREAKER/$25–$90

MACHINE GUNNER— ROCK 'N ROLL/$25–$90

BAZOOKA SOLDIER—ZAP/
$30–$140

COUNTER INTELLIGENCE—
SCARLETT/$35–$220

LASER RIFLE TROOPER—
FLASH/$25–$90

also by their code names. The figures came with backpacks and weapons, and each had a command dossier with many details about the fig-

ure including personal data and military specialization.

Each of the figures in the first assortment, produced in 1982, had

a straight arm with only an elbow joint, which caused diffi-culty in posing the fig-ures with its weapons. These first nine figures came packaged on

COBRA OFFICER/$25–$130

COBRA COMMANDER/
$25–$110

COBRA/$25–$140

MEDIC—DOC/$25–$75

S.E.A.L.—TORPEDO/$25–$75

**MINE DETECTOR—TRIPWIRE/
$25–$75**

cards that listed the military classification of the figure much more prominently than the figure's code name.

Infantry Trooper Grunt, Ranger Stalker, Mortar Soldier Short Fuze, Communications Officer Breaker, Machine Gunner Rock 'n Roll and Bazooka

Soldier Zap all appear fairly similar and are dressed in military and khaki green. The black figure of Commando Snake Eyes and the female

MARINE—GUNG-HO/$25–$75

**ARCTIC TROOPER—SNOW JOB/
$25–$75**

**HELICOPTER ASSAULT
TROOPER—AIRBORNE/
$25–$75**

figure Counter Intelligence Officer Scarlet are twice as valuable than the other first series figures and can fetch around $200 in mint condition in its original packaging.

Laser Rifle Trooper Flash, whose military green uniform has red accents, is the only figure of the series to feature brightly colored attire.

The new line of G.I. Joes was dedicated to fighting against the aggressive forces of the enemy, Cobra. Cobra Officer was offered later, along with a Cobra Com-

MAJOR BLUDD/$20–$60

DESTRO/$35–$140

FIRST SERGEANT—DUKE/$15–$50

HEAVY MACHINE GUNNER—ROADBLOCK/$35–$85

TRACKER—SPIRIT/$20–$65

DOG HANDLER—MUTT & JUNKYARD/$20–$65

mander figure that was featured as a mail-in promotion. The value of the Cobra Commander figure varies depending on if it is straight-armed or a later version.

A new series of figures featuring a "swivel arm battle grip" was designed to solve the problem posed by the straight-arm figures, and was manufactured in 1983.

Included along with all the original nine were a stretcher-bearing Medic figure named Doc, a Navy S.E.A.L. named Torpedo, Mine Detector Tripwire, Helicopter Assault

COBRA INTELLIGENCE OFFICER —BARONESS/$40–$175

COBRA SABOTEUR—FIREFLY/ $40–$90

COBRA NINJA—STORM SHADOW/$40–$160

FIRST SERGEANT—DUKE/ $40–$80

ZARTAN/$55–$105

**INFANTRY TROOPER—
FOOTLOOSE/$15–$45**

**SILENT WEAPONS—
QUICK KICK/$15–$45**

The enemy forces were increased and included new Cobra villains like Major Bludd and metal-masked Destro. Duke, the G.I. Joe leader, was offered as a mail-in. The silverheaded Destro figure proved to be the most popular villain produced in the 3¾" G.I. Joe line.

**COVERT OPERATIONS—
LADY JAYNE/$25–$100**

**COMMANDO—SNAKE EYES II/
$35–$140**

The third series, released in 1984, included even more new Cobra enemy figures such as Anti-Armor Specialist Scrap Iron, the bespectacled Intelligence Officer Baroness, Saboteur Firefly, and the white-clad Ninja Storm Shadow. The figures of the third series were more creative and varied in appearance than those of the first two series. The bright orange-and-red-clad Flame Thrower Blow Torch figure certainly

Trooper Airborne, and a tattooed Marine named Gung-Ho. Arctic Trooper, part of the same series, is reminiscent of the original 12" Ski Patrol

figure, with its white ski suit and white skis. In this series, the figure's code name was printed on the package more prominently than the classification.

stands out from the mostly neutral colored figures of the first and second series. Tracker Spirit is based on a Native American Indian, and Dog Handler Junkyard comes with Mutt. Heavy Machine Gunner Roadblock, Halo Jumper Rip Cord, Jungle Trooper Recondo, Zartan with Swamp Skier, and First Sergeant Duke rounded out the third series of figures, which also included an expanded offering of attack vehicles and playsets. A Hooded Cobra Commander figure was featured as a mail-in promotion.

The fourth series of figures continued to diversify the line—including more women and minority figures—and to vary the figures' uniforms further. Along with relatively traditional-

SAILOR—SHIPWRECK/$25–100

WARRANT OFFICER—FLINT/ $25–$100

COBRA FROGMAN—EELS/ $25–$75

COBRA POLAR ASSAULT— SNOW SERPENT/$25–$75

looking figures like Infantry Trooper Footloose and Warrant Officer Flint, the series included figures of a martial artist called Quick Kick, a female

covert operations figure called Lady Jayne, Sailor Shipwreck with parrot, and an updated Commando figure named Snake Eyes with a timber

CRIMSON GUARD COMMANDERS— TOMAX & XAMOT/$35–$95

ICEBERG/$10–$35

SCI-FI/$10–$35

wolf. Also part of the series are the colorful figures of Airtight, a Hostile Environment operative, and Fire Fighter Barbecue. New Cobra figures in

the series included Frogman Eels, Polar Assault Snow Serpent, and the Crimson Guard Commanders, twins Tomax and Xamot. A mail-in Sgt.

Slaughter figure was also available.

By the fifth series, released in 1985, the figures had become even more colorful,

ZANDAR/$8–$30

ZARANA/$8–$30

LEATHERNECK/$10–$36

WET-SUIT/$15–$50

HAWK/$18–$50

B.A.T./$18–$50

detailed and varied. Snow Trooper Iceberg wore a nicely detailed white suit, and Laser Trooper Sci-Fi wore a bright green suit with silver boots and trim.

The redheaded brother/sister enemy team of Zandar and Zarana were dressed in multicolored neon clothing. Two versions of Zarana were pro-

duced—the facial features of the original, more valuable figure are larger and the eyebrows are more pronounced.

DR. MINDBENDER/$10–$35

COBRA COMMANDER/$18–$45

CRYSTAL BALL/$8–$15

Leatherneck, Lifeline, Wet-Suit, Roadblock, and Hawk were versions of standard Joe character types, but the elaborate enemy figures B.A.T., a Cobra Android Trooper, and Dr. Mindbender, a Master of Mind Control, represented new types of Joe figures. A unique mail-in offer figure of The Fridge, based on Chicago Bears football player William "Refrigerator" Perry, was also manufactured at this time.

By the sixth series, most of the figures were unrecognizable when compared with traditional G.I. Joe figures. New, complex,

BIG BOA/$10–$35

JINX/$10–$38

RAPTOR/$10–$30

CROC MASTER/$10–$35

GUNG-HO (DRESS UNIFORM)/ $15–$40

BLOCKER/$10–$40

and intriguing enemies like Cobra Commander with Battle Armor, Cobra Trainer Big Boa, menacing Cobra Hypnotist Crystal Ball, Falconer Raptor and Reptile Trainer Croc Master certainly had no counterparts in the original G.I. Joe series. Novelist and G.I. Joe fan, Stephen King, wrote the bio information for the Crystal Ball character card back. The female heroine Jinx came dressed in a hot-pink ninja outfit with accessories. However, a more traditional Gung-Ho figure in U.S. Marine dress uniform was also part of the series.

AVALANCHE/$10–$40

MAVERICK/$10–$40

DODGER/$10–$40

SHOCKWAVE/$8–$35

BUDO/$8–$35

VOLTAR/$6–$25

Specialized vehicle drivers like Blocker, Avalanche, Maverick, and Dodger were part of the 1987 Battle Force 2000 Special Forces set. The hi-tech figures were all operators of military vehicles like hovercrafts or half-tracks and came packaged on new cards with a streamlined Battle Force 2000 logo. The Battle Force 2000 vehicles were sold separately.

The seventh series included new figures like S.W.A.T. team member Shockwave and Samurai Warrior Budo, as well as new Battle Force 2000 figure Voltar with a condor, advertised as one of the Iron Grenadiers. Spearhead and his bobcat Max, a black-and-white Storm Shadow II figure, and Serpentor were also manufactured at this time. The popular

SPEARHEAD & MAX/$6–$23

STORM SHADOW II/$15–$40

HOODED COBRA COMMANDER/$20–$40

Steel Brigade figure could be named and customized by the child as his or her own personalized G.I. Joe figure.

SGT. SLAUGHTER/$15–$35

The eighth series mixed both new figures like Scoop and Iron Grenadier T.A.R.G.A.T. with updated versions of popular older figures

THE FRIDGE/$15–$30

SERPENTOR/$20–$40

SNAKE EYES III/$18–$38

T.A.R.G.A.T./$8–$20

GNAWGAHYDE/$8–$20

like Snake Eyes III and Rock 'n Roll II. A Dreadnok figure named Gnawgahyde came packaged with a warthog and had a handlebar mustache, and leopard print

accents on his uniform. A Slaughter's Marauders assortment with a new Sergeant Slaughter figure was manufactured with a new logo and revised packaging.

The now massive line continued to expand with a ninth and then a tenth series released in 1990 and 1991. Mail-in orders continued to be offered to children and collectors—Life-

LIFELINE/$6–$15

RED STAR/$12–$23

DESTRO II/$7–$12

line was part of a Kellogg's Rice Krispies promotion. The eleventh series Red Star figure came on two different cards, a regular card and a more valuable Cobra card. The figures in the eleventh series had action features and offered new versions of popular characters like Destro II with a shooting Disc Launcher along with cannon-shooting Wild Bill. The line continued to multiply through the thirteenth series in 1993.

WILD BILL/$4–$9

EFFECTS/$10–$25

LOBOTOMAXX/$15–$28

PREDACON/$25–$45

CARCASS/$20–$40

**30TH ANNIVERSARY
ACTION SOLDIER/$10–$20**

The fourteenth and final regular series is the most difficult to understand because of variations in the order of production, and planned figures

that were never produced. In 1994 the series was released on new borderless cards with the logo placed vertically on the card rather than horizon-

tally. The series was divided into different assortments like Star Brigade and Battle Ninja. The Star Brigade figures Lobotomaxx, Predacon, and Carcass

**30TH ANNIVERSARY
ACTION SAILOR/$10–$25**

**30TH ANNIVERSARY
ACTION MARINE/$10–$20**

**30TH ANNIVERSARY
ACTION PILOT/$10–$28**

30TH ANNIVERSARY ORIGINAL ACTION TEAM/$30–$75

were highly detailed, with monstrous features and elaborate pictures on the packaging. Countdown, Ozone and Effects were similarly detailed with intricate accessories.

Even though the line was already discontin-ued, a set of 3¾" scale figures manufactured to commemorate the 30th anniversary of G.I. Joe, was made. It included Action Soldier, Action Sailor, Action Marine, and Action Pilot along with an Action Set with all four figures and an exclusive astronaut figure and Mercury space capsule. The packaging in the individual figures was reminiscent of the original G.I. Joe figures and duplicated their original uniforms and accessories.

HALL OF FAME 12″ **G.I. JOES**

HASBRO, 1991–94

TARGET DUKE
COBRA COMMANDER
SNAKE-EYES
DESTRO
STORM SHADOW

The Hall of Fame series was the result of a collaboration between Hasbro and the Target chain of stores. The first figure, Duke, was a Target exclusive, which was later released in an altered mass-market version. The figure was heavily promoted and when released in stores sold out within an hour. The figure used a body mold

TARGET DUKE/$10–$35

COBRA COMMANDER/$10–$30

SNAKE-EYES/$12–$35

from a doll in the Maxie line of toys and added a head made to look like the head of the 3¾" Duke. Each figure was individually numbered and came with a sticker affixed on the figure's back just above the waist.

The success of the figures resulted in a series return of 12" G.I. Joe figures. Hasbro was very conservative in their production and release because of high production costs, although demand for them

turned out to be much stronger than expected.

The popular and long-lasting figures of Cobra Commander,

Snake-Eyes, Gung-Ho, Destro, Storm Shadow, Flint and Rock 'n Roll each appeared in yet another incarnation.

DESTRO/$25–$40

STORM SHADOW/$20–$35

LIMITED EDITION 30TH ANNIVERSARY G.I. JOES

HASBRO, 1994
SIZE: 12"

30TH ANNIVERSARY
ACTION SOLDIER

30TH ANNIVERSARY
ACTION SAILOR

30TH ANNIVERSARY
ACTION MARINE

30TH ANNIVERSARY
ACTION PILOT

A new 12" body closer to the 1964 G.I. Joe was created for the 30th anniversary of G.I. Joe in 1994. Action Soldier, Action Sailor, Action Pilot, and Action Marine figures were rereleased once again with copies of their original outfits and packaging and were well received.

**30TH ANNIVERSARY
ACTION SOLDIER/
$40-$200**

**30TH ANNIVERSARY
ACTION SAILOR/
$50-$220**

**30TH ANNIVERSARY
ACTION MARINE/
$40-$200**

**30TH ANNIVERSARY
ACTION PILOT/
$60-$240**

G.I. JOE— SGT. SAVAGE AND HIS SCREAMING EAGLES

HASBRO, 1994–95
SIZE: 6"
VALUE RANGE: $4–$10

COMMANDO SERGEANT SAVAGE

The proliferation of 6" G.I. Joe figures contin- ued in 1995 after the cancellation of the huge "A Real Ameri- can Hero" line. The line included 12 fig- ures along with vehi- cles and accessories, and was built around a Sergeant Savage character. The first Commando Sergeant Savage came with a 22-minute animated video relating the ori- gin of this new Joe team. The figures came packaged on cards with a new Sergeant Savage logo and each figure had

COMMANDO SERGEANT SAVAGE

more and better accessories than pre- viously provided.

GOBOTS

TONKA, 1985–86
SIZE: MULTI-SCALES
VALUE RANGE: $2–$8

FLIP TOP

It seems appropriate that a toy company known for making toy trucks would intro- duce a line of robots that transformed into vehicles with pull-back action. The GoBots, along with Hasbro's Transformers, were tremendous successes in the mid-1980s.

The GoBots changed into cars, trucks, heli- copters, jeeps, tanks— you name it! Each GoBot came blister- packed on a card with a picture of the vehicle into which it transformed, and a holographic sticker. Flip Top changed into a helicopter.

FLIP TOP

GODZILLA

MATTEL, 1977
SIZE: 19½"
VALUE RANGE: $50–$250

GODZILLA

This 19½" Godzilla was manufactured as part of Mattel's Shogun Warriors line. He has a "flame" tongue and firing hands and looks a good deal like the creature that appears in so many Japanese monster movies. Although hundreds of versions of Godzilla have been made in Japan, only a few were imported to the United States. This Godzilla is one of the better, more "realistic" toys.

GODZILLA

GODZILLA, KING OF THE MONSTERS

TRENDMASTERS, 1994–95
SIZE: 4" AND 10"
VALUE RANGE: $2–$15

GODZILLA

Godzilla action figures made a comeback in the mid-1990s with the release of the Trendmasters series. This line featured monsters in both 4" and 10" sizes, plus most of Godzilla's adversaries. This Godzilla has "glowing atomic plasma eyes." Each figure comes with a collectible trading card. The Godzilla Force was also part of this line. They were four human figures, each with heavy arsenals, ready to fight the monsters.

GODZILLA

GODZILLA WARS

TRENDMASTERS, 1996
SIZE: 4" AND 10"
VALUE RANGE: $4–$12

GODZILLA

This line was essentially a continuation of the Godzilla, King of the Monsters series. Old figures were repackaged in window boxes. Snap-on armor and weapons were also added. The artwork on the box was more menacing than the pictures in the previous series. These figures also came with collectible trading cards.

GODZILLA

GOLDEN GIRL AND THE GUARDIANS OF THE GEMSTONES

GALOOB, 1984–87
SIZE: 6"
VALUE RANGE: $5–$18

GOLDEN GIRL
VULTURA

These fully poseable figures were hybrids— part action figure, part fashion doll. Each figure had action features and accessories, but doll-style clothing outfits with names like "Festival Spirit" and

GOLDEN GIRL

"Forest Fantasy" were also available. Some outfits even included purses.

VULTURA

Golden Girl and the other "good" characters were named after gemstones and precious metals. Vultura and the other evil figures were named after insects and monsters. Both good and evil female figures are quite Amazonian but also have long flowing hair that somehow reasserts their femininity. All are packaged with reversible cloth capes, a die-cast gemstone, a weapon and a weapon belt, and a comb. Heaven forbid these women should have a bad-hair day when they go into battle.

The Golden Girl line also had two male figures—Prince Kroma and Ogra. Ironically, these figures are more valuable to collectors than the female figures.

THE GREATEST AMERICAN HERO

MEGO, 1981
SIZE: 3¾"
VALUE RANGE: $75–$175

CONVERTIBLE "BUG" WITH RALPH AND BILL

This convertible Volkswagen "bug" comes with 3¾" action figures Ralph and Bill, the two main characters of the humorous science-fiction television series **The Greatest American Hero.** William Katt and Robert Culp starred in the story of hapless high school teacher Ralph Hinkley (Katt),

CONVERTIBLE "BUG" WITH RALPH AND BILL

who is given a magic suit by aliens so he can help save the Earth from self-destruction. He is aided by Bill Maxwell, an FBI agent also chosen by the aliens. The show aired from 1981 to 1983, and the main character's last name

was changed from "Hinkley" to "Hanley" after John W. Hinckley, Jr.'s, assassination attempt on President Ronald Reagan.

Three 8" figures are advertised on the back of the box but were never produced.

HAPPY DAYS

MEGO, 1978
SIZE: 8"
VALUE RANGE: $8–$50

FONZIE

Arthur Fonzarelli aka Fonzie aka "the Fonz" was one of the most popular characters of popular culture in the mid to late 1970s. Although he was not originally positioned to be the star of the hit television show **Happy Days**, it became clear after only a few episodes that

Richie Cunningham's ultra-cool greaser buddy with a heart of gold was a cultural phenomenon.

The Fonzie action figure features Fonzie's trademark leather jacket, jeans and white T-shirt, but perhaps the cleverest feature is the "thumbs-up" action. On the TV show, Fonzie's sign that all was cool was his exclamation "eyyyy!" accompanied by a double "thumbs-up."

FONZIE

This could be imitated on the figure by positioning Fonzie's arms and hands, and positioning his thumbs in the "up" position, and then pressing the button on his back. Eyyyy!

HONEY WEST

GILBERT, 1965
SIZE: 12"
VALUE RANGE: $85–$300

HONEY WEST

Honey West was a popular detective drama that aired on ABC from September 1965 to September 1966. The character of Honey West had first appeared on an

episode of **Burke's Law.** This Honey, TV's "new girl P.I.," as played by Anne Francis, was a sexy judo and karate expert devoted to her pet ocelot, Bruce; her partner, Sam; and her aunt Meg. She had inherited her detective business from her father. Many baby boomers will remem-

HONEY WEST

ber her specially modified lipstick that contained a radio transmitter, and the cosmetic compact that had a hidden camera in its mirror. She was most often dressed in a black leo-tard, with a gold belt and black boots.

The action figure, billed as "TV's private eye-full," came dressed in the black judo leotard. Her accessories included a gold belt and gun holster, a pistol, and of course, the black boots. She looked a lot like Anne Francis, right down to the Marilyn Monroe–like beauty mark just to the right of her luscious lips.

THE CRASH DUMMIES

TYCO, 1991–94
SIZE: 5"
VALUE RANGE: $5–$15

VINCE
LARRY

Perhaps what was most incredible about The Crash Dummies is that an educational-toy series devoted to teaching safety was so successful. These figures were originally based on the Department of Transportation's (DOT) TV mascots, the crash dummies Vince and Larry, whose purpose was to remind people to wear seat belts. Their slogan, "you

VINCE

could learn a lot from a dummy" appeared on each card. Their message was greatly enhanced by these action figures, since they "explode" or fly apart on impact.

After a few months, the DOT withdrew its support because of complaints about

LARRY

small parts. But Tyco had a success on their hands. They simply re-named the main characters "Slick" and "Spin" and reprinted cards that had no references to Vince, Larry, and the DOT. The early cards are highly collectible.

INDIANA JONES

KENNER, 1982–83
SIZE: 12"
VALUE RANGE: $10–$235

INDIANA JONES

This 12" Indiana Jones figure from Kenner was made in conjunction with the Adventures of Indiana Jones line. Both were based on the movie blockbuster **Raiders of the Lost Ark**. This wonderfully styled figure came with removable cloth hat, jacket, shirt, pants, boots, holster, gun and whip. He was window-boxed with two photographs of Harrison Ford as he appears in the movie.

INDIANA JONES

INDIANA JONES AND THE TEMPLE OF DOOM

LJN, 1984
SIZE: 6"
VALUE RANGE: $40–$115

INDIANA JONES

Another Harrison-Ford-as-Indiana-Jones action figure; this time taken from the second Indiana Jones movie, **Indiana Jones and the Temple of Doom**.

A fully articulated 6" "Indy" is dressed again in his trademark outfit. His accessories include a whip, a very big knife, an explorer's bag and the character's trademark fedora. This short-lived series is hard to find. Other figures in the series included Mola Ram and Giant Thuggee. Short Round and Willie Scott are advertised on the back of the blister cards, but

INDIANA JONES

none were ever produced.

INHUMANOIDS

HASBRO, 1986
SIZE: 14"

AUGER (6½")
REDLEN
MAGNOKOR
METLAR

The Inhumanoids was a mid-1980s TV animated series about a group of California earth scientists—the Earth Corps—and their battles with the Inhumanoids, evil beings mutated from rocks or redwood trees. One wonders what the environmental message was behind this—evil redwood trees?

Dr. Edward Agutter, or Auger, as he is known by friends, is a short-tempered ex-boxer who became a distinguished archaeologist and joined the Earth Corps. He was responsible for the construction of Earth Corps environmental suits and weapons and their vehicles, the Terrascout and the Trappeur.

AUGER/$5–$15

Redlen is one of the mutores—a mutant redwood tree with menace on his old-

MAGNOKOR/$10–$70

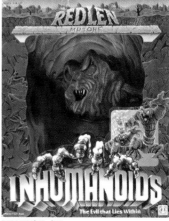

REDLEN/$10–$70

138

growth brain. Magnokor, another mutore made of solid rock, is a creature from Earth's core. He is capable of splitting into two separate polarized beings: One half, Pyre, is fiery and has a temper to match; the other half, Crygen, is icy and fueled by no passion. Pyre and Crygen have the capacity of either attracting or rejecting each other, depending on how they are aligned. They can position themselves on either side of Metlar, Magnokor's archenemy, and paralyze him using their reverse polarity. Metlar is the ultimate enemy. He comes from Earth's mantle and battles Inhumanoid and human alike. He is a fierce opponent to all he battles.

All of the Earth Corps human action figures

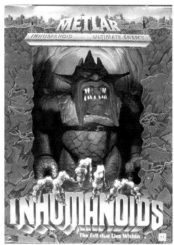

METLAR/$10–$70

come with a removable glow-in-the-dark helmet and an excavation tool. Auger comes dressed in an environmental suit and carries a spinning auger drill. The human figures are blister-packed on cards that show the Inhumanoids logo, and a cross section of Earth.

Redlen, Magnokor, and Metlar are large 14" figures. The

window-box design makes it appear that they are emerging from a cave. Redlen is collapsible and has glowing eyes; Magnokor can be separated into his two parts, Pyre and Crygen; and Metlar has glowing fangs and horns. A merchandise catalogue was included with each figure.

INSPECTOR GADGET

TIGER, 1992–94
SIZE: 5¼"
VALUE RANGE: $5–$20

**INSPECTOR GADGET
DR. CLAW**

The European animated television show **Inspector Gadget** was a hit with kids from the minute it first appeared in syndication in North America in the mid-1980s. Galoob first immortalized the bumbling private investigator (who had the voice of **Get Smart**'s Don Adams aka "Maxwell Smart") with a single action figure in 1984, but when **Gadget** hit the big time and began airing on Nickelodeon, Tiger came out with an entire series of Inspector Gadget action figures.

There were six versions of the basic 5¼"

INSPECTOR GADGET

DR. CLAW

Gadget action figure, each with a different color trench coat and set of accessories. The Inspector pictured here wears a gray coat and comes with a snap-open map and treasure-hunting tools. All Inspector Gadget figures are blister-packed on cards, with a color picture of the Inspector; his niece, Penny; and her dog, the Brain. Under Gadget's feet appears his trademark slogan "Wowsers!"

The Tiger series also featured the Inspector's archenemy, Dr.

Claw, the leader of the evil organization M.A.D. Although his face was never seen on the television show, it was revealed for the first time on this action figure. The figure was blister-packed with a small sign placed in front of Dr. Claw's face, so his visage was not visible unless you bought the figure and opened the package. Dr. Claw comes with a dropping bomb, a top secret file, and his pet, M.A.D. Cat. M.A.D. Cat is pictured in color on the card.

IRON MAN

TOY BIZ, 1995–96
SIZE: 5"

**IRON MAN
SPIDER-WOMAN
U.S. AGENT**

The Iron Man series was introduced when the animated **Marvel Action Hour** debuted in 1994. These figures are wonderfully detailed and painted, with many action accessories and bold, colorful packaging.

There are several versions of Iron Man, including a 10" figure. The Iron Man pictured here comes with Iron Man armor and a plasma cannon missile launcher. Other versions of the character included one with deep-sea weapons and one with hologram armor and a power missile launcher. The 10" Iron Man wore space

IRON MAN/$5–$15

SPIDER-WOMAN/$5–$15

armor and carried a laser weapon.

Spider-Woman, one of the female Marvel Super Heroes, came with "psionic web hurling action." She wears a purple suit that has a white spider on the chest. She is also packaged with a biography card. Iron Man was not.

The U.S. Agent figure was canceled when retailers demanded smaller case assortments. Only about 100 bagged figures of U.S. Agent were pro-

U.S. AGENT/$300–$600

duced, and most of them were given to Toy Biz employees. A few have made their way into the hands of collectors. Needless to say, they are very valuable.

Spider-Woman and Iron Man came blister-packed on a card with a picture of Iron Man. The name of the character appeared in typeface designed to look like a green LED display. All the smaller packages have a sticker on the lower-right corner of the blister that has the Marvel logo with the words "as seen on the Marvel Action Hour."

JAMES BOND

GILBERT, 1965–66
SIZE: 12"
VALUE RANGE: $100–$400

JAMES BOND
ODD JOB

Fresh on the heels of **Thunderball, Dr. No,** and **Goldfinger,** Gilbert released some very cool 12" James Bond action figures. These figures were carefully designed with great attention to detail in dress, facial expressions and accessories. They are excellent additions to any collection.

The James Bond figure, very much resembling Sean Connery, is dressed in a cloth trench coat and is based on the Bond

JAMES BOND

ODD JOB

from the movie **Thunderball.** His accessories include binoculars, numerous disguises, a S.C.U.B.A. outfit complete with snorkel, fins, bathing trunks, and a mask, and a cap-firing gun that can be triggered by a spring-action arm.

Odd Job, the silent but deadly killer from **Goldfinger,** is dressed

in a cloth martial arts robe with a black belt. He is packaged with his deadly derby, which can be launched by a spring-action arm. He also has karate-chop action.

Both figures are packaged in cardboard boxes with color pictures of the character as they appear in the films.

JURASSIC PARK

KENNER, 1993–94
SIZE: 5" (DIMENSIONS
VARY)
VALUE RANGE: $16–$50

TYRANNOSAURUS REX
VELOCIRAPTOR

The quality of the action dinosaurs from the Jurassic Park series is truly spectacular, and it is what ensured the line's success. Not only did kids recognize the quality and want the "real" thing, Kenner stamped each character and dinosaur with a "JP brand" to indicate that it was official Jurassic Park merchandise, and the packages warned "if it's not Jurassic Park™, it's extinct!"

The detailed dinosaurs could roar, stomp, bite and spit. Major dinosaurs, like the Tyrannosaurus Rex, had rubber "dinosaur skin" stretched over a plastic frame, with all the action mecha-

TYRANNOSAURUS REX

nisms and joints concealed. This made it much more realistic. The T-Rex also came with electronic roar and stomp sounds and a limited-edition movie collector card. It is window-boxed, with the movie logo on the box. The figure is placed against a jungle backdrop.

The smaller 5¼" Velociraptor figure is

no less detailed. Though he does not have the rubber "skin" of the larger figures, he has the "JP" brand on his haunches, is realistically painted, and has "Dino-Strike" slashing jaws. He is blister-packed on a card with the Jurassic Park jungle background and logo, and a movie collector card is included.

VELOCIRAPTOR

KNIGHT RIDER

KENNER, 1983–86
SIZE: 6"
VALUE RANGE: $35–$125

**KNIGHT 2000 VOICE CAR
WITH MICHAEL KNIGHT**

KNIGHT 2000 VOICE CAR WITH MICHAEL KNIGHT

Long before David Hasselhoff found fame and fortune with **Baywatch,** he played Michael Knight, second banana to a souped-up, talking 1982 black Pontiac Trans Am named K.I.T.T. on the hit action television series **Knight Rider.** Knight and the car, voiced by William Daniels, investigated crimes outside the jurisdiction of the police.

The battery-powered Kenner "Knight 2000 Voice Car," which came with a Michael Knight action figure, was a replica of the mechanical TV star. It had the authentic K.I.T.T. computer voice and could say five different phrases. On the box was a photograph of the toy with the action figure and an inset photograph of David Hasselhoff as Michael Knight.

KOJAK

EXCEL TOY CORP., 1976
SIZE: 8"
VALUE RANGE: $50–$150

KOJAK

The action figure with the "who-loves-ya, baby" attitude, Kojak not only came with glasses and a police revolver and holster, but he's also accessorized with lollipops and cigars, a fedora, and a cloth pin-stripe suit. Based on the Telly Savalas hit television show, the Kojak action figure is fully articulated. Although his face isn't quite Telly, his bald head is a perfect match. He is blister-packed on a card with plain type.

KOJAK

THE LEGEND OF THE LONE RANGER

GABRIEL, 1982
SIZE: 4"
VALUE RANGE: $5–$15

THE LONE RANGER

The Legend of the Lone Ranger series was based upon the 1981 movie **The Legend of the Lone Ranger.** The film was an attempt to revive the Lone Ranger franchise and to present actor Klinton Spilsbury as the new masked rider of the Plains. The hope was that this film would be the first of a number of Lone Ranger big-screen adventures. But audiences did not cotton to it, and the film, and its star, became a footnote in the history of movies based on serials, comic books, and television and radio series.

The action figures were nicely done, however. The Lone Ranger figure wears his trademark black mask and white Western outfit and hat, and is posed at the ready to draw his pistols. (Gabriel ran a special mail-in offer with the figures—if you bought

THE LONE RANGER

four Lone Ranger figures or horses, you would get a Western town free.) All figures are blister-packed on a card, with the movie's logo and a picture of the Lone Ranger's face.

LEGENDS OF BATMAN

KENNER, 1994–96
SIZE: 6½"
VALUE RANGE: $5–$25

CYBORG BATMAN
CATWOMAN
THE LAUGHING MAN JOKER
DARK RIDER BATMAN

This series was based on the DC Comics **Elseworlds** comic books and other alternative versions of Batman. **Elseworlds** heroes, according to DC Comics, "are taken from their usual settings and put into strange times and places—some that have existed, and others that can't, couldn't or shouldn't exist. The result is stories that make characters who are as familiar as yesterday seem as fresh as tomorrow."

CYBORG BATMAN

CATWOMAN

THE LAUGHING MAN JOKER

The action figures of Legends of Batman reflect this "familiar originality." There is no doubt the figures are of the well-known Batman, Catwoman, and Joker, but they are presented in revitalized ways. All figures are of high quality and detail, as are the best Kenner figures. But Batman, in this series, is bigger and more muscular. Cyborg Batman seems to be inspired by the Borg on

Star Trek. This is not a traditional Batman. He comes with a "laser" weapon, and his cybernetic eye lights up. His paint indicates that he is part human, part cyborg.

Catwoman looks especially fierce with her long black gloves, black hip boots and whip. She wears her mask, but her bright red lips and long black hair are visible. She also comes with a

grappling hook and a capture net. The Laughing Man Joker is dressed as a mad pirate, complete with purple pirate hat and waistcoat, green hair and pants, and huge buckle. He is part of the "pirate special edition" in this series. Resembling a technicolor Captain Hook, the Joker grins maniacally and wields powerful "Gatling gun attack."

All three of these figures are blister-packed and come with an official collector's card with a picture of the specific character in action. The Legends of Batman logo, with a picture of Batman, is on the upper-left corner of the card.

Dark Rider Batman comes in a box with a color picture of Batman riding a stallion. The enclosed Batman figure sits astride a "rearing battle stal-

DARK RIDER BATMAN

lion." This Batman comes with a whip and a sword. He has

"whipping arm and slashing sword action."

THE LION KING

MATTEL/ARCO, 1994
SIZE: 4"
VALUE RANGE: $3–$9

**FIGHTING ACTION
ADULT SIMBA
COMIC ACTION PUMBAA
WITH TIMON**

The enormously successful Disney animated feature **The**

Lion King inspired a fanciful and well-made line of action figures for younger children. These moveable figures with push-button action are sculpted and decorated to look exactly like their movie counterparts.

Adult Simba comes with "fighting action." He is blister-packed on a card bearing a color background from the movie and a picture of Simba. It appears that the figure is standing in front of Pride Rock. Comic Action Pumbaa, the hysterical warthog, can open and close his jaws. He is packaged with his sidekick, the meerkat Timon. Both are painted grinning. Timon and Pumbaa are blister-packed on a card with a color background of the jungle. There is a picture of Pumbaa as he appears in the movie.

FIGHTING ACTION ADULT SIMBA

COMIC ACTION PUMBAA WITH TIMON

LITTLE DRACULA

DREAMWORKS, 1991
SIZE: 4"
VALUE RANGE: $10–$30

LITTLE DRACULA

This series of high-quality, imaginative figures was released in time for Halloween in 1991. If there can be such a thing as a "cute" vampire, Little Dracula was it. The action figure was available in two models— one with a cape that unfurled at the touch of a button and another with "light up, electric eyes" (battery included!). Both came with "vampire gear" and their pet bat, "Batty." All the figures are bubble-packed on cards with a color picture of a cartoonish Dracula's castle. The Dreamworks Toy Company of 1991 is not related to the Dreamworks SKG Film Company.

LITTLE DRACULA

THE LITTLE MERMAID

TYCO, 1989–93
SIZE: 11½"
VALUE RANGE: $20–$40

ARIEL (SHELL ON PACKAGE)

The Little Mermaid is credited with revitalizing the Disney animated feature genre. It was brilliantly written, directed and cast, and it was animated by the best artists in the business. The story of the plucky mermaid who wanted more out of her life touched a chord with children and their parents. It was tremendously successful and gave birth to a line of wonderful toys. Ariel influenced other doll makers; there are several mermaid-themed dolls on the toy-store shelves.

This Ariel was the first Little Mermaid figure to be available with the release of the movie. She wears a removable glittery green fish tail, a purple bikini top and has long flowing red hair. She is window-boxed, and pictures of her friends Flounder and Sebastian, the hermit crab, appear to be looking at the doll.

ARIEL

There is also a picture of Ariel as she appears in the movie.

THE LONE RANGER RIDES AGAIN!

GABRIEL, 1979
SIZE: 12"
VALUE RANGE: $10–$45

THE LONE RANGER
TONTO
BUTCH CAVENDISH

The "daring and resourceful Masked Rider of the Plains" and his "faithful Indian Companion" Tonto returned in 1979 to fight for law and order against the head of the Hole-in-the-Wall Gang, Butch Cavendish, who had murdered his brother.

These 12" action figures are almost a throwback to the early days of action figures, most notably

THE LONE RANGER

TONTO

BUTCH CAVENDISH

the Best of the West line. The Lone Ranger figures have a similar attention to detail and theme as these earlier lines. Each figure came with a small

comic book and was packaged in a four-color box with a picture of the enclosed character in an action pose. All figures were fully jointed and accessorized with guns,

cloth outfits and other accessories. Horses for the figures were also available, as were adventure packs with themed accessories.

LOONEY TUNES

TYCO, 1994–95
SIZE: 8"
VALUE RANGE: $3–$9

BUGS BUNNY
TASMANIAN DEVIL
MARVIN THE MARTIAN

In 1994, Tyco picked up the Looney Tunes license and released a

new line of figures with action features. The larger talking figures are downright adorable, in spite of their violent tendencies. They are carefully painted and sculpted to accurately reflect the joy and mania of the 'toons.

Bugs Bunny, smart-alecky as ever, comes with a "carrot missile" and launcher. The Tasmanian Devil has "super spin tornado twist action," barbecue tools, and a hot dog on a skewer. Tiny Marvin the Martian has a huge weapon

BUGS BUNNY

TASMANIAN DEVIL

MARVIN THE MARTIAN

with a "plutonium 200 missile." All three are blister-packed on cards with a picture of

the character and the Looney Tunes logo. Each came with a Looney Tunes trading

card from Pyramid, a trading card company.

LORD OF THE RINGS

KNICKERBOCKER, 1979
SIZE: 5⅛" (DIMENSIONS VARY)
VALUE RANGE: $15–$225

GANDALF THE GREY

ARAGORN

FRODO

GOLLUM

SAMWISE

RINGWRAITH THE BLACK RIDER

The classic J.R.R. Tolkien trilogy, **The Lord of the Rings**, was the basis for Saul Zaentz's 1979 animated feature film. The Knickerbocker figures based on the movie are prized by collectors for their amazing detail, gorgeous paint job and fluid sculpting. The figures were based on

the film model sheets, with the exception of the Ringwraith and his charger.

Gandalf the Grey, the wizard, is 5⅛". He wears long flowing blue robes, a removable wizard's hat, a wizard's staff, and has a long white beard. His wizened face

continued on page 154

The Toy Vault:
Lord of the Rings' Figures Are Born

John Huston, a big J.R.R. Tolkien fan, decided he wanted to have **Lord of the Rings** action figures created that were true to the book descriptions. He approached several toy companies with the idea. None were interested. So he set upon a journey much like Frodo's. He acquired the rights and sourced the means to produce the figures himself.

John Huston studied the paintings and sculpture of fantasy artists he felt could best help him realize his dream. He made his choice and worked with the sculptors to get them exactly the way he had envisioned. Once the first prototype was created, he went to Hong Kong to interview toy manufacturers and discuss his project with several different companies. With his prime choices in mind he returned to the U.S. to discuss package design, marketing and promotion.

The result of his hard work was a new toy company which John named Toy Vault. This new venture would produce the Tolkien figures, two at a time, under the

The Fire Balrog

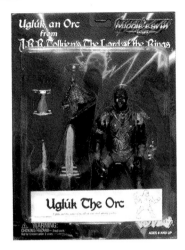

Uglúk The Orc

banner of Middle Earth Toys. The dream of a fan turned into a full-time business. John's company successfully marketed the first six figures in the line and has four more in the process of production and marketing at this writing. The collector base and demand for action figures is now large enough for individuals with vision, who with hard work, are able to realize dreams not before possible.

Gandalf the Wizard

GANDALF THE GREY

ARAGORN

FRODO

shows his great age, while his bright blue eyes demonstrate his great wisdom. Until he was lost in the dark depths of Moria, he was part of the Fellowship of the Ring and told the Hobbits of the true nature of the Ring.

Aragorn, a great hero, is 5" high. He wears a belted green tunic and brown boots and comes with a sword and scabbard. His careworn face shows courage and nobility. He was blessed with three times the normal life span of a human.

Aragorn traveled widely and fought for the rights of free peoples in his quest to become King of Amor and Gondor. After Gandalf was lost in Moria, Aragorn became the leader of the Fellowship of the Ring, and eventually King.

Frodo, the "star" of the tale, is a 3" hobbit. His light brown tunic and dark brown trousers are atypical of a Hobbit—they prefer to wear bright yellows and greens. But, like all Hobbits, Frodo is half the size of a human, with hair on

his bare feet. He carries a sword and scabbard. Frodo inherited the Ring of power from his uncle. Years later, he learned from Gandalf that the Ring was evil Sauron's and that it must be destroyed. Frodo's quest to carry the Ring to Mordor and destroy it is the main tale of **The Lord of the Rings**.

Gollum is an obsequious little creature. He is a previous owner of the Ring and wants it back. The power of the Ring and his greed reduced him to a slithering dark

SAMWISE

GOLLUM

RINGWRAITH THE BLACK RIDER

hairless wraith, but it is believed he was once a Hobbit—he still has Hobbit feet. He is 3" and completely dark gray. He wears a black loincloth.

Samwise is Frodo's friend and manservant. The figure is 3" and carries a sword and scabbard. Samwise accompanies Frodo on his quest and is loyal to the very end.

Ringwraith the Black Rider is the most desirable of all these figures. In the movie, these creatures were

presented as shadowy, mostly live-action images with little definition, so Knickerbocker's artists had to create a menacing dark rider every bit as fearful as the one on the screen. They succeeded handily. This 5¼" figure is evil incarnate. He wears a great black hooded cloak, a black chain-mail shirt, and a horned helmet. To give the impression of having no body, the figure has no face—just glowing red eyes in a field of darkness. Ringwraiths are the mightiest and most

powerful of all Sauron the Ring Lord's evil servants. Weapons could not harm a Ringwraith unless it was magically forged by the elves, and any other blade that touched a Ringwraith would perish.

All the Lord of the Rings action figures are blister-packed on cards, with a color picture of the character against a pale mustard-colored background. The character's name and the title of the movie is set in stylized Gothic-looking type.

THE LOVE BOAT

MEGO, 1981
SIZE: 3¾"
VALUE RANGE: $5–$15

CAPTAIN STUBING

The popular Aaron Spelling television extravaganza **The Love Boat** was a big hit in the 1980s. So popular that Mego decided to create Love Boat action figures. Never mind that most kids didn't stay up to watch the show on Saturday nights or that the series was decid-edly not action-oriented, the figures appeared in stores in 1981. A large replica of the ship was planned, but never made. Captain Merrill Stubing was "your ship's captain" and was played by actor Gavin MacLeod. The figure wears the captain's uniform: a short white jacket with gold buttons and black trousers. He comes with a captain's white hat, which is removable. He is blister-packed on a card with

CAPTAIN STUBING

a photograph of the **Pacific Princess** in the background. A cast photo inside a life preserver shows all the figures in the series.

THE MAD MONSTER SERIES

MEGO, 1974
SIZE: 8"
VALUE RANGE: $25–$185

DRACULA
FRANKENSTEIN
WOLFMAN
MUMMY

Even though they are not heroes, Mego released the Mad Monsters as part of their Official World's Greatest Super Heroes line. These 8" articulated action figures are based on classic Universal movie monsters, and the painting and rendering closely reflect them. All figures have glow-in-the-dark hands and eyes, and are packaged in cardboard boxes with the Mego Super Heroes logo at the top, next to the Mad Monster logo.

Dracula wears his white shirt and black

waistcoat with a red, satin-lined black cape. His black hair meets his forehead in a widow's peak, and his face is as white as his shirt. He looks more than a little like Bela Lugosi. He comes in a cardboard box, with a color painting of "the dreadful Dracula." A few head variations are known to exist. Some Dracula figures have red hair, a different face, and a red vest and tie painted onto the white shirt.

Frankenstein has a flat head with black hair and gray skin. He wears a black jacket and dark gray shirt. His face resembles Boris Karloff's makeup in the classic film. His box has a painting of "the monster Frankenstein" against a fiery orange background. Frankenstein is also found in a variation with blue hair and a different face.

DRACULA

FRANKENSTEIN

WOLFMAN

MUMMY

The Wolfman's paint is mottled and camouflaged in variations of brown and gray to resemble fur. He has a brown mane of hair on his head, as well as fur down the bridge of his nose and on his chin. He has brown furry hands and feet. The picture on his box is a truly frightening painting of "the human Wolfman."

The Mummy is painted in shades of brown and tan, and is sculpted to look as if he is wrapped with bandages. He has a visible human face with some bandaging on it, and light brown hair. His box has a painting of "the horrible Mummy."

MAGNUM P.I.

LJN, 1985
SIZE: 4½"
VALUE RANGE: $25–$85

CAR AND ACTION FIGURE

In the 1980s, everyone's favorite TV private investigator was Thomas Magnum, a Vietnam vet who lived in Hawaii. Tom Selleck played the part with a twinkle in his eye— Magnum was a caring but tough shamus who was very happy to be practicing his craft on his island paradise. This Magnum figure comes with

CAR AND ACTION FIGURE

Magnum's signature red Ferrari.

The Magnum action figure really doesn't do justice to Selleck's good looks. The mustachioed figure wears the same casual blue jeans and denim shirt, but the toy's face can't capture the lighthearted spirit of the Magnum character. The Ferrari is a classic—very red and very suave, this plastic replica is a great toy. The figure can be placed in the driver's seat.

Magnum and his Ferrari are window-boxed with a photograph of Tom Selleck against a Hawaiian sunset on the cardboard flap. Another smaller photo is at the lower-right area of the plastic window and shows Magnum leaning against his car.

MAJOR MATT MASON

MATTEL, 1966-70
SIZE: 6"

MAJOR MATT MASON
CALLISTO
SCORPIO

In 1966, when the NASA Gemini-and Apollo-manned space programs were capturing the imagination of the American public, Mattel released Major Matt Mason: Mattel's Man in Space. According to Mattel, all of Major Matt Mason's equipment, including his space suit, was based on official space program designs.

The figures are wire-reinforced rubber "flexies" with accordion joints. These serve the same purpose as articulated joints on molded plastic action figures, so the Major and his colleagues were, in fact, action figures. The fun of the Major Matt Mason figures is in the details and variety of the characters and accessories. By buying just one figure, a child would have an astronaut and many different play options. The Major Matt line was a rich experience indeed.

MAJOR MATT MASON
$60–$220
Major Matt Mason was the only figure available during the first year of this series. He was sold on a blister card with a set of accessories (later called the "Flight Set") or with a "Moon Suit."

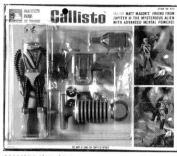

CALLISTO/$75–$350

SCORPIO/$500–$1,400

According to Mattel, Major Matt Mason's Moon Suit gave him "even more protection" than his space flight suit from "extreme temperature changes and radiation while exploring the planet and collecting samples." Also included with the Moon Suit figure were tools, "two launch packs with which Major Matt Mason can launch rockets and space probes into hard-to-reach areas," a Space Sled, which "moves him across the moon's surface," and a short range "Jet Propulsion Pack." Major Matt's space helmet was removable. It came with an attachable radiation shield.

Callisto, an alien, was made in 1969. This green-headed figure was Major Matt's friend from Jupiter who had "advanced mental powers." He was blister-packed with a purple "Space Sensor" that mounted on his shoulder. It could fire and retract a plastic line. Scorpio, another alien produced in 1970, was the Major's "mysterious ally from the stars." He is a very mod pink and purple, and looks a little "Star Trek–ish." He came window-boxed with a bellows-controlled vest that fired "search globes." He has removable arm and leg shields, and the head has a small light inside.

THE MAN FROM U.N.C.L.E

GILBERT, 1965
SIZE: 12"
VALUE RANGE: $75 – $250

NAPOLEON SOLO
ILLYA KURYAKIN

For nearly four years (Sept. 22, 1964, to Jan. 15, 1968), millions tuned in to the adventures of agents Napoleon Solo (Robert Vaughn) and Russian agent Illya Kuryakin (David McCallum), who represented the United Network Command for Law and Enforcement. U.N.C.L.E.'s mission was to use their wits and lots and lots of cool gadgets to rid the world of the diabolical organization Technical Hierarchy for the Removal of Undesirables and the Subjugation of Humanity (T.H.R.U.S.H.). If the show seemed James Bond–like, it was no accident. Author Ian Fleming was involved

NAPOLEON SOLO

ILLYA KURYAKIN

in the early stages of developing the series. He even contributed the names of the agents.

The U.N.C.L.E. action figures are of the fine quality and detail of early-sixties figures, especially the figures from Gilbert. Napoleon Solo and Illya Kuryakin both came in cardboard boxes with a photo of the character on the front. Each figure was equipped with a real

cap-firing gun, which was triggered by releasing his spring-loaded arm. U.N.C.L.E. pocket insignias were included, as well as an U.N.C.L.E. ID card. Illya wears his trademark black turtleneck, while the suave Solo wears his tuxedo. Several action figure accessory packs could be purchased separately. For example, one included many kinds of guns— bazookas, pistols, rifles, and a telescope.

MARVEL SUPER HEROES

TOY BIZ, 1990–94
SIZE: ???⅛"
VALUE RANGE: $5–$25
(EXCEPT DAREDEVIL)

SPIDER-MAN
CAPTAIN AMERICA
DAREDEVIL
DR. DOOM
DR. OCTOPUS
THE INCREDIBLE HULK
SILVER SURFER
THE AMAZING SPIDER-MAN WITH REAL WEB-SHOOTING ACTION
TONY STARK/IRON MAN
GREEN GOBLIN
THOR
VENOM
MR. FANTASTIC
THING
INVISIBLE WOMAN
HUMAN TORCH

SPIDER-MAN

In 1990, super heroes from the Marvel Comics universe were included in the Toy Biz line for the first time. There was a heavy demand for the figures, but Marvel was not satisfied with the sculpting of the characters. There were a number of shipping delays, and demand could not always be met. Since Marvel wanted to gain an expanded presence on toy-store shelves, they ultimately achieved this by buying a 46% interest in Toy Biz in 1993.

As a result, Marvel essentially gained a controlling interest in Toy Biz. With Marvel controlling the production and design of the figures, quality was greatly improved. Since Marvel could now decide which figures would be pro-

CAPTAIN AMERICA

DAREDEVIL/$20–$60

DR. DOOM

duced, **Marvel Super Heroes** and its spin-off lines produced a wide variety of figures, including many minor characters that other companies would have omitted.

Perhaps most famous of the Marvel characters is Spider-Man. The first Spider-Man in the Toy Biz line has "web-suction hands." The small suction cups on his fingers can adhere to smooth surfaces. Spider-Man wears his traditional blue and red outfit and comes blister-packed with a black plastic spider-web. There is a color

picture of the Marvel comic-book character on the card.

Captain America is one of the heroes of Marvel's "Golden Age." This figure wears Captain America's red, white, and blue suit, and winged mask. This figure is blister-packed on a card with a picture of the comic-book character and comes with a shield and shield launcher.

Daredevil looks a lot like Spider-Man, but in an all-red suit. He has the same body as both Spider-Man and Captain America, with

a different paint job. Daredevil comes with an "extending billy club." Like the other figures in this series, he is blister-packed on a card with a picture of the comic-book character. This figure is the most valuable of the first-year figures.

Dr. Doom is another of the 1990 figures, the first year of the series. He wears his trademark green cape and hood. His limbs are silver, and his black-belted tunic is green. He comes with "power driven" weapons. He is known as the enemy

163

of the Fantastic Four, but also creates a lot of problems for other Marvel Super Heroes.

Dr. Octopus is a fun figure of a bad guy. He wears his geeky glasses and green suit and has very long clear plastic mechanical arms with suction cups on the end. Originally created to safely manipulate toxic materials from a distance, Dr. Octopus became physically and mentally fused to these arms after a freak industrial accident. The formerly mild-mannered Otto Octavius became the infamous insane criminal and enemy of Spider-Man known as Dr. Octopus.

The Incredible Hulk figure is bubble-packed on a card with the "Hulk" logo larger than the Marvel logo, and a color picture of the comic-book char-

DR. OCTOPUS

THE INCREDIBLE HULK

SILVER SURFER

THE AMAZING SPIDER-MAN

acter. He is bright green with purple pants and comes with a steel bar that he can bend with his "crushing arm action." The Incredible Hulk, of course, is the muscle-bound alter ego of Dr. Robert Bruce Banner. He was created when

Dr. Banner was exposed to gamma radiation. It caused him to mutate into a powerful "hulking" physical manifestation of his inner struggles.

The Silver Surfer is perhaps the most elegant of the Marvel Super Heroes, and

TONY STARK/IRON MAN

GREEN GOBLIN

THOR

VENOM

the line. Its body is more finely sculpted and its paint job more sophisticated. It comes with a black spiderweb, which it can shoot with a web-shooting attachment. Spider-Man is bubble-packed on a full-color card, with a color picture of Spider-Man and the Spider-Man logo once again displayed more prominently than the Marvel logo.

Tony Stark/Iron Man is another example of the finer figures that Toy Biz created once Marvel had a controlling interest in the company. This figure is a terrific representation of the Vietnam vet, Tony Stark, and his alter ego, Iron Man. Iron Man comes with "quick change armor," which includes body armor and a red and yellow mask. The figure is bubble-

his action figure aptly reflects this. He is a lean muscular figure painted completely silver and comes with a silver "action" surfboard. The Silver Surfer, also known as Norrin Radd, was born in the Milky Way

and is an interstellar explorer.

The Amazing Spider-Man with "Web Shooting" action is a slightly more elaborate figure than the one with web-suction hands. This figure reflects the Marvel influence over packed on a full-color card with the Iron Man logo prominently displayed above the Marvel logo.

The Green Goblin, Spider-Man's archenemy, is an insane genius with super powers. This figure comes with "pumpkin bombs," which can be launched by pressing the button on his back. This figure is carded on a color card with the Green Goblin logo appearing prominently above the Marvel logo. The background on this card is the actual comic strip, in black and white. All the figures in this series have their comics on their card as a motif.

Muscle-bound Thor, the Norse god of thunder, comes with long blond flowing Nordic hair and a horned helmet. Dressed in his two-

MR. FANTASTIC

toned blue armor, wrist bands, and golden-strapped boots, this is a very nicely detailed figure. Thor comes with a hammer and "smashing hammer action," as well as thunderbolts that he can hurl from the sky.

The Venom figures were the most difficult to find. Shortly after Venom with "living skin slime pores" was released, it was replaced with a similar figure that squirted "alien liquid" (a.k.a. water). Venom is one of the coolest figures,

THING

with an all-black bodysuit and mask. A white arachnid graces his chest, and white wings cover his eyes.

The Fantastic Four were part of the Marvel Super Heroes "Cosmic Defenders" line. When a rocket ship on a mission into deep space was struck with heavy cosmic rays, scientist Reed Richards, his fiancée Sue Storm, her brother Johnny Storm, and Reed's friend Ben Grimm absorbed massive radiation, which mutated them into

INVISIBLE WOMAN

HUMAN TORCH

beings with incredible powers.

Mr. Fantastic, Reed Richards, has the ability to stretch his body into all kinds of bizarre forms. His action figure has "Five-way stretch." He is dressed in a blue bodysuit with the "4" logo emblazoned on his chest.

Ben Grimm was transformed into the Thing, a living rock pile. He has super strength and endurance but cannot change back into his human form. This action figure

shows his mass well. He is covered with golden "rocks" and is clearly not human, although he is painted with a gentle human smile. He comes with "pulverizing punch," and a street sign that he can bend as he wishes.

Sue Storm gained the ability to become invisible at will. The color-changing Invisible Woman action figure had very limited distribution and became one of the most famous short-run figures ever produced. The same mold was

used for another Invisible Woman in 1994, but it did not have color-changing paint. A launcher arm was added to the 1994 figure.

Sue's brother Johnny Storm was changed into the Human Torch, a super hero with the ability to control fire and create it in his body. He can fly, too. This figure is quite spectacular, with vibrant red and yellow paint. It has "fireball flinging action" and comes with three "fireballs." All of the Fantastic Four come bubble-packed on full-color cards, with the Cosmic Defenders boxed-in logo in the upper left-hand corner of the card. The Fantastic Four logo is the most prominent brand name on these cards, with the names of the individual characters in the lower-left part of the card.

MARVEL SUPER HEROES SECRET WARS

MATTEL, 1984
SIZE: 5"
VALUE RANGE: $5–$70

CAPTAIN AMERICA

SPIDER-MAN

IRON MAN

WOLVERINE

DOCTOR OCTOPUS

SPIDER-MAN, BLACK COSTUME

HOBGOBLIN

ICEMAN

CONSTRICTOR

ELECTRO

In this 1984 Mattel series, Marvel Super Heroes Secret Wars, the most popular Marvel Super Heroes were suddenly transported to a remote planet known as the "Beyonder." The heroes battled with the most powerful villains of the time. This event was chronicled in a twelve-part special comic-book series and later in regular comics. This Marvel series also

CAPTAIN AMERICA

SPIDER-MAN

IRON MAN

WOLVERINE

loaned itself to many other toys and products.

These action figures were developed by Mattel at a time when they were still riding high on the success of

168

DOCTOR OCTOPUS   **SPIDER-MAN, BLACK COSTUME**   **HOBGOBLIN**

ICEMAN   **CONSTRICTOR**   **ELECTRO**

their **Masters of the Universe** action figure line. They developed a sleeker, different look for these Marvel

characters. This series also marked the first time Marvel characters had an action-figure

line based solely on Marvel characters. In the **Captain Action** line and the Mego's **Official World's**

Greatest Super Heroes series, Spider-Man and Captain America had appeared with DC characters.

Each of the Marvel Super Heroes in this series came with a "secret shield" that had four different convex card inserts.

Sixteen figures in all were produced. Thirteen figures were sold on English cards in the United States. Iceman, Constrictor and Electro were sold on French and Spanish cards in Canada, South America and Europe. Each card has the Marvel Super Heroes Secret Wars logo against a dark sky-blue background, with a lightning bolt.

Each figure wears his trademark Marvel outfit. Each character's secret shield has a color picture of the character's face. Each character comes with his trademark weapon.

M*A*S*H

TRI-STAR INTERNATIONAL, 1970, 1982–83
SIZE: 12"

HAWKEYE
HOT LIPS
HAWKEYE (3¾")

The classic Robert Altman film that wove together thousands of dark and sardonic quips into a single compelling narrative, was eventually spun off into one of the most popular sitcoms in television history. M*A*S*H was as much a part of the cultural zeitgeist in the 1970s and early 1980s as **Seinfeld** was in the 1990s.

Hoping to capitalize on the show's incredible popularity, Tri-Star International produced two sets of M*A*S*H action figures. They were not terribly successful, but one must remember that the show was not

HAWKEYE/$15–$35

HOT LIPS/$15-$35

geared toward children.

Prepubescent boys, however, may have been attracted to the 12" Hot Lips figure. This replica of the sexy head nurse of the 4077 came dressed in removable white scrubs with fatigues underneath, combat boots and a hat. She is bubble-packed with a medkit on a card with black-and-white photographs from the show. Twelve-inch Hawkeye Pierce comes dressed the same as Hot Lips. He is also bubble-packed on the same card.

The 3¾" M*A*S*H figures offered a wider choice. All of the major characters are represented in this series. The packaging is zippier, with a photograph of the famous helicopter bringing wounded on the top of the card with the M*A*S*H logo superimposed upon it. The lower part of the card shows a color photograph of the entire cast. Above each character is a dog tag with the character's name, and to the right of the figure is the road sign from the show.

HAWKEYE/$5-$15

These figures seem more animated than the larger ones. The sculpting is quite good for smaller figures. All are smiling and dressed in army fatigues with the exception of two. Winchester wears a white medical coat, and one of the two Klingers wears a pink dress. Klinger in a dress and Hot Lips were the most difficult to find.

THE MASK

HASBRO/KENNER, 1995–96
SIZE: 5″
VALUE RANGE: $5–$15
(EXCEPT TALKING MASK)

WILD WOLF MASK
BELLY BUSTIN' MASK
CHOMPIN' MILO
TALKING MASK (14″)

Jim Carrey's manic performances as the guy who went "from zero to hero" in the big-screen hit **The Mask** was more cartoonish than most and seemed perfectly suited for adaptation as a Saturday morning TV series. Kids loved the movie, they loved Carrey and they loved the cartoon. It followed that kids would love action figures based on **The Mask**, and they did.

The Mask action figure line was released in the midst of Hasbro's acquisition of Kenner in 1995. The toys were

WILD WOLF MASK

BELLY BUSTIN' MASK

CHOMPIN' MILO

designed by Hasbro's creative staff and sold by Kenner. The first series packages are labeled "Hasbro Toy," but subsequent figures and accessories are labeled "Kenner."

The colorful toys were designed with both boys and girls in mind. Each figure featured stunts that the movie and cartoon characters had performed. Wild Wolf Mask came with a giant "punching" bazooka, a cartoonish wolf face, and

is dressed in the same bright yellow suit Carrey wears in the movie. Belly Bustin' Mask was rumored to be canceled because of concern over the "squirting Milo" packaged with it, but it was sold in very limited quantities in some markets. Belly Bustin' has an exploding stomach and is dressed in a blue suit and tie. The enclosed "squirting Milo" does just that. Chompin' Milo, the masked hero's dog, has a big green face with huge teeth—a result of the dog's trying on the Mask. He comes with a bone that he can chomp on and a shooting net launcher strapped onto his back. All three come bubble-packed on a card with a photo of the movie character in the upper part of the card and a picture from the cartoon on the lower part.

The rarest figure in this series is the Talking Mask. Only 500 of this

TALKING MASK/$100–$800

larger figure were produced. It is fully poseable and utters famous phrases, like "sss-smokin'," from the movie and series when a button is pushed.

MASTERS OF THE UNIVERSE

MATTEL, 1981–1990
SIZE: 6″
VALUE RANGE: $5–$20

BATTLE ARMOR HE-MAN

BATTLE ARMOR SKELETOR

BEAST MAN

BUZZ-OFF

BUZZ-SAW HORDAK

DRAGSTOR

EVIL-LYN

EXTENDAR

FAKER

FISTO

HE-MAN

HURRICANE HORDAK

JITSU

KING HISS

KING RANDOR

KOBRA KHAN

LEECH

MAN-AT-ARMS

MAN-E-FACES

MANTENNA

MEKANECK

MER-MAN

MOSQUITOR

MOSS MAN

NINJOR

ORKO

PRINCE ADAM

RAM MAN

ROKKON

ROTAR

SAUROD

SCARE GLOW

SKELETOR

SNAKE FACE

SNOUT SPOUT

SORCERESS

SPIKOR

SSSQUEEZE

STINKOR

STONEDAR

STRATOS

SY-KLONE

TEELA

TRAP JAW

TRI-KLOPS

TWISTOID

TUNG LASHOR

TWO BAD

WEBSTOR

WHIPLASH

ZODAC

BATTLE ARMOR HE-MAN

BATTLE ARMOR SKELETOR

BEAST MAN

At the height of the Masters of the Universe's popularity in the mid-1980s, Tom Wolfe used the term "masters of the universe" in his book **The Bonfire of the Vanities**. The bond traders on Wall Street in this 1980s-defining novel referred to themselves this way. They were young, incredibly wealthy and invincible—much like He-Man, the "most powerful man in the universe."

Just as the social mores of the 1980s were elucidated by the Wolfe book, the toy business was just as much influenced by the successful line of Mattel action figures. Star Wars and G.I. Joe had marked milestones in the history of action figures—the Masters of the Universe was just as much of a turning point in the toy world as these previous lines.

The Masters of the Universe line was the first line of action figures that incorporated action features. After that, toy manufacturers tried to include some kind of action feature, even if it was no more than an accessory that snapped onto the figure.

More significant to the success of the line was a Federal Communications Commission (FCC) ruling in December 1983. It was a real Christmas gift for tele-

BUZZ-OFF

BUZZ-SAW HORDAK

DRAGSTOR

175

vision programmers and toy manufacturers alike. The FCC lifted a number of restrictions on children's television programming. Among them was a 1969 law prohibiting television shows based on children's toys.

Filmation Associates was quick to act on the freer atmosphere of deregulation. With the cooperation of Mattel, they produced sixty-five half-hour animated episodes of **He-Man and the Masters of the Universe**. The major networks weren't interested, so Filmation traded the animated sequences for a portion of the airtime and allowed local stations to keep the advertising revenues. So Mattel achieved what no other toy company had before—a weekly 30-minute commercial

EVIL-LYN

EXTENDAR

FAKER

FISTO

for their toy line directed precisely at its captive target audience. The success of this marketing campaign was not lost on

other toy manufacturers, and soon product-based TV shows and videos were created for toy lines such as The Care

HE-MAN

HURRICANE HORDAK

JITSU

KING HISS

their marketing staff had decided Conan was too violent and sexual for children's toys. So, Conan became He-Man, and Mattel introduced Masters of the Universe in 1981.

All figures in this series are bubble-packed on cards with the Masters of the Universe logo at the top and the character name underneath.

The hero of the series is Prince Adam the leader of Eternia. He possesses a magic sword which transforms him into He-Man, the "most powerful man in the universe" when he says "I have the power!" The original He-Man action figure and Battle Armor He-Man are extremely muscular. They are both sculpted with small waists to empha-

Bears, Transformers, Strawberry Shortcake and countless others. The trend continues today.

But what of the toys themselves? In the early 1980s, Mattel had planned a line of Conan the Barbarian action figures, but

KING RANDOR

KOBRA KHAN

LEECH

size their large arms and legs. The heads are identical, with blond, Prince Valiant-type hair and knee-high sandals. Original He-Man's waist can be twisted to throw a power punch. Battle Armor He-Man has a breastplate that shows a dent when struck and can be repaired by a touch of the hand. Both figures come with the power sword.

Skeletor is He-Man's archenemy, and the

ruler of Eternia's sister planet, Infinita. He is leader of an army set on taking control of He-Man's fortress, Castle Grayskull. The Skeletor action figure and Battle Armor Skeletor both have a hooded, eyeless skull head, and carry a staff. Like the original He-Man, Skeletor can throw a punch when his waist is twisted, and Battle Armor Skeletor comes with the same kind of repairable armor. Skeletor's body is identical

to He-Man's, but his arms and legs are painted an icy blue.

Beast Man is one of Skeletor's henchmen. The orange-bodied creature with a simian face throws a power punch when its waist is twisted, and it carries a weapon.

Buzz-Off, the "heroic spy in the sky," is part bee, part human. The figure's torso is striped black and yellow, and the figure has a yellow face with in-

MAN-AT-ARMS

MAN-E-FACES

MANTENNA

sect eyes. Buzz-Off has poseable wings, which can be extended by twisting his waist.

Buzz-Saw Hordak is the leader of the Evil Horde and former teacher of Skeletor. This group of enemies was introduced mid-series and was packaged with comic books. The Evil Horde are creatures who fight both He-Man and Skeletor. Buzz-Saw Hordak has a saw-blade that bursts from

his chest when his waist is twisted.

Dragstor, another member of the Evil Horde, is a transforming toy. This heavily armored figure changes into an "evil warrior vehicle." Possibly using Hot Wheels technology, another Mattel line, Dragstor has ripcord action, which propels the vehicle.

Evil-Lyn is in cahoots with Skeletor. She is the evil warrior goddess and bears a

passing resemblance to Xena. Like most female action figures, she is slightly anatomically exaggerated and is dressed in what appears to be a blue-armored bathing suit. She has bright yellow skin and returns a punch if you twist her waist. She comes with a weapon.

Extendar, the "heroic master of extension," is heavily armored in silver and gold. His arms and legs can be extended to make him the "tallest heroic warrior ever."

The original Faker is one of the more valuable figures. The figure was reissued. He is basically He-Man's evil robotic twin. Faker was created by Skeletor. The figure is a duplicate of the He-Man figure, but his skin is the same blue as Skeletor's, and he has red hair.

Fisto is He-Man's "heroic hand-to-hand fighter." The bearded figure has an extra large right fist which throws a punch when his waist is twisted. His torso is armored.

Hurricane Hordak is one of the more interesting figures of the series. The figure of the ruthless leader of the Evil Horde comes with "wicked whirling weapons." This Hordak has an action "ro-

botic" arm that wields a variety of interchangeable weapons.

Jitsu is the evil counterpart to Fisto. One

MEKANECK

MOSQUITOR

of Skeletor's army, he is the "evil master of martial arts." This bearded figure throws karate chops with his oversized right arm

MER-MAN

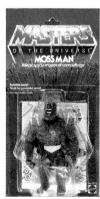
MOSS MAN

when his waist is twisted. He wears red and gold armor.

King Hiss is the "dreadful disguised leader of the Snake Men." This figure transforms into a mass of serpents. The crowned figure's skin has a snakeskin texture.

King Randor, He-Man's father, is the heroic ruler of Eternia. This figure has the same body as He-Man, dressed in red and gold armor, but an older-looking head. He is bearded and wears a crown.

Kobra Khan is the "evil master of snakes" and not to be confused with King Hiss. This scary figure sprays "venom" or water when his head is pressed down. He has a green reptilian body and snake eyes.

With a name like "Leech," he has to be bad, and he is. Leech is the "evil master of power-suction." This figure, another member of the Evil Horde, is bulkier than most of the figures in this se-

NINJOR

ORKO

PRINCE ADAM

RAM MAN

ries. He has a dark green body and wears black armor with a bat on his breast. His face is one large suction cup, with teeth, that sticks to smooth surfaces.

Man-At-Arms is the heavily armored master of weapons. This green-skinned figure has the power-punch action of the other early figures in this series.

Man-E-Faces is an "heroic human . . . robot . . . monster." This very cool figure has those three different faces on a rotating head.

Mantenna, an evil spy for the Evil Horde, has a blue insectlike body and head, and wears the Horde's red bat symbol on his chest. His eyes pop out when his waist is twisted.

Mekaneck, the heroic human periscope, has a periscope neck that pops up when his waist is twisted.

Mer-Man is the undersea counterpart to He-Man. He is an ocean warlord. The black figure has a stylized diving helmet for a head and wears a crown. He throws a power punch when his waist is twisted.

The Mosquitor figure, part of the Evil Horde,

ROKKON

ROTAR

SAUROD

is an "evil energy-draining insectoid." This figure has a red head with a long stinger. A red fluid in his chest actually pumps through his body.

Moss Man, a "heroic spy and master of camouflage," has a dark green textured body. This figure comes with the power-punch action feature.

Ninjor tapped into the appeal of the Ninja theme in action-figures. He is an "evil Ninja warrior" and comes with special "Ninja weapons." He can wield them by twisting his waist. He is dressed in a black ninja jumpsuit and hood. His eyes are visible.

Orko is He-Man's court magician. He once saved He-Man's life. This action figure spins across the floor

when it is wound up. Also included is a magic trick.

Prince Adam is He-Man's "heroic secret identity." He is He-Man without the sword of power. This figure is painted in a white shirt and vest, "civilian clothes," but in all other ways is an exact duplicate of the He-Man figure.

Ram Man is a "heroic battering ram." This heavily armored figure

SCARE GLOW

SKELETOR

SNAKE FACE

has a spring-loaded body that can be pushed down and released so it can "batter" against the walls of enemy strongholds like Snake Mountain, the lair of Skeletor and the evil warriors.

Rokkon, the "young heroic comet warrior," transforms from meteor to mighty warrior. The figure is light blue with a "comet" helmet.

Rotar, "heroic master of hyper-spin," has super gyro spin action. This figure is half human, half-toy top. He comes with a gyroscope that can be used to propel him into a spinning action. Twistoid is his evil counterpart. Both toys come with weapons they can wield while spinning.

Saurod, the "evil spark-shooting reptile,"

SNOUT SPOUT

SORCERESS

SPIKOR

SSSQUEEZE

shoots real sparks from his mouth. This figure is brown with a reptilian face. On the package, the **Masters of the Universe** full-length motion picture is advertised underneath the figure.

STINKOR

STONEDAR

STRATOS

SY-KLONE

Scare Glow is a nifty glow-in-the-dark figure. He is the evil ghost of Skeletor. The figure is purple, but for its skeleton, which is made of luminous plastic that glows in the dark after exposure to a light source.

Snake Face, the "most gruesome of the Snake Men warriors," has a gray face with bright red eyes and fangs. He has snakes on his torso. When his waist is twisted, snakes come popping out of his chest and face.

Snout Spout, an "heroic water-blasting firefighter," has the standard He-Man body painted orange, and an elephant head with a long trunk. This figure can shoot water, or a "jetspray," from his snout.

The guardian of Castle Grayskull is the Sorceress. She keeps watch over Eternia, flying under the guise of Zoar the falcon. She uses her mystical powers to defend the

TEELA

TRAP JAW

TRI-KLOPS

side of good. It is the Sorceress who told Prince Adam his destiny and gave him the sword of power. This female figure comes dressed in a white leotard and carries a magic staff. She has wings that spring open.

Spikor is the "untouchable master of evil combat." This figure's entire body resembles a mace, with spikes all over the dark blue torso and head. His right arm can tele-

scope out to wield his weapon.

SSSqueeze is an "evil long-armed viper." This figure has long arms that look like snakes and can constrict around his victims, in this case, other action figures. He is bright green with a reptilian face.

Stinkor, "evil master of odors," comes in orange armor and really does have an odor. "It's fun" says

the package. He also has the power-punch feature.

Stonedar, "heroic leader of the Rock People," can transform into a warrior from a boulder and vice versa. This light blue figure can be "folded" into a boulder.

Stratos, a winged warrior, is He-Man's gray flying friend. He lives in the mountaintop city of Avion with his fellow Birdmen. This figure has a falcon

TWISTOID

TUNG LASHOR

TWO BAD

head, and blue wing attachments on his arms. He also has the power-punch feature. Another version of Stratos had red wings.

Sy-Klone is He-Man's "heroic fist-flinging tornado friend." This light blue and bright yellow figure has an animated radar screen on his torso. His torso can be spun 360 degrees.

Teela, stepdaughter of Man-At-Arms, one of the guards of Castle

Grayskull, is the real daughter of the Sorceress, although she does not know it. She is a skilled warrior who often helps He-Man. The action figure wears a white-and-gold armored suit with snake decorations. She has the power-punch action feature.

Trap Jaw is "evil and armed for combat." Trap Jaw has the ability to change his many weapons by detach-

ing and reattaching them from his arm. He comes heavily armored and armed, but the highlight of this figure is the "free" warrior's ring that comes packaged with it. The ring is kid-size.

Tri-Klops, the "evil warrior who sees everything," also comes with a warrior's ring and power punch. Tri-Klops has three eyes, as his name suggests.

WEBSTOR

WHIPLASH

ZODAC

Tung Lashor is an "evil tongue-shooting snake creature." This bright purple figure with orange torso has tongue-lashing action. His forked snake tongue "lashes out" of his snake mouth two inches when his waist is turned.

Two Bad is a "double-headed evil strategist." This figure has two fanged heads—

one purple and one blue. Its body is painted with the right side blue and the left side purple. When its waist is twisted, the arms form a bear hug.

Webstor, "evil master of escape," is a dark blue figure with a Spider-Man–like head. He comes with a line of rope that he can climb when his waist is twisted.

Whiplash is an "evil tail-thrashing warrior." This bright chartreuse figure has an armored torso much like a turtle shell. His large tail thrashes when his waist is twisted.

Zodac is the "cosmic enforcer." This figure is masked and comes with power-punch action.

MAXIMUM CARNAGE

TOY BIZ, 1994
SIZE: 5"
VALUE RANGE: $5–$20

CARNAGE

This line was a one-shot tie-in with the release of the video game of the same name. Most of the Marvel-based figures in the series were multipacks—Carnage with

Spider-Man, for example. Only Carnage came packaged alone, as well. The single-packaged Carnage came in two paint variations. This figure is red with black markings and has either white or sharply defined teeth outlined in black. The Carnage figure with outlined teeth is slightly more valuable. A bonus collector's pin

CARNAGE

is included with the blister-packed figure.

McDONALDLAND CHARACTERS

REMCO, 1976
SIZE: 8"
VALUE RANGE: $15–$60

RONALD McDONALD
BIG MAC
HAMBURGLAR
MAYOR McCHEESE
CAPTAIN CROOK
GRIMACE

In the early seventies, McDonald's introduced several promotional characters who live in McDonaldland with Ronald McDonald, the fast-food

RONALD McDONALD

chain's mascot clown. Television commercials showed a live-action, Syd and Marty Kroft–style set, with actors in costumes playing

BIG MAC

Mayor McCheese, the Hamburglar and others.

The popularity of these characters surged with each TV

HAMBURGLAR

MAYOR McCHEESE

CAPTAIN CROOK

GRIMACE

mark yellow and orange. Each character is packaged on a pedestal with the name of the character on the front. All the toys are fully poseable and have moveable plastic heads.

Unlike most action figures, the McDonaldland characters come dressed in cloth costumes. Ronald McDonald wears his clown suit, Big Mac wears a trench coat and a belt with the McDonald's logo on the buckle, the Hamburglar wears a cape and tie over his striped prison outfit, Mayor McCheese wears a pin-striped suit and his mayoral sash, and Captain Crook is dressed in his pirate duds and carries a scabbard. Grimace bears a slight resemblance to the Cookie Monster. He is purple and soft—a squeezable cuddly toy.

ad, and in 1976, Remco manufactured a line of McDonaldland action figures based on the stars of the ads. Remco is known primarily for producing generic figures based on best-selling toy lines. But the McDonaldland figures were original, had tons of personality and were imaginatively designed. Each figure came blister-packed on a full-color card. There is a picture of all the citizens of McDonaldland against a background of McDonald's trade-

MICRONAUTS

MEGO, 1977–79
SIZE: 3¾", 5", 8"
VALUE RANGE: $5–$40

TIME TRAVELER
SPACE GLIDER
GALACTIC WARRIOR
ACROYEAR
ACROYEAR II
PHAROID
BIOTRON
FORCE COMMANDER
OBERON
BARON KARZA
ANDROMEDA
MEGAS
REPTO
MEMBROS
ANTRON
GIANT ACROYEAR
HORNETROID

The very popular Micronauts line was the Americanized version of the Japanese Micro Man series by Takara. The major feature that set Micronauts apart from other action figures was the five-millimeter holes and matching five-millimeter posts incorporated into each figure.

TIME TRAVELER

SPACE GLIDER

The result was a fully interchangeable line of toys. Every Micronaut item could be assembled in various ways, by itself, and in hundreds of combinations with other Micronaut toys. Items were made from die-cast metal and/or plastic. Some incorporated spring-load missiles, which were later banned.

Mego marketed the toys aggressively. Television commercials emphasized the inter- changeability of the items by asking "What can you make?" Commercials ran on both network and local television stations, aimed at both children and their parents. Some toys came blister-packed on a four-color card, and some came in full-color boxes. In Mego's promotional materials, they included color banners for stores, blowups of the Micronaut logo and items. Some stores set up

GALACTIC WARRIOR

Micronaut "centers" so customers could see all of the toys on display and the variations that could be made.

There were four different varieties of figures. Each was fully articulated and poseable and was scaled for use with the Micronaut play sets and vehicles. They came in assorted colors and had science-fiction–sounding names.

ACROYEAR

ACROYEAR II

PHAROID

BIOTRON

FORCE COMMANDER

Time Traveler was the basic Micronaut. He's transparent and is the basic building block for the interchangeable world of the Micronauts. Space Glider is a flying Micronaut. He's die-cast and comes with snap-on wings that open and shut by pressing a button. Galactic Warrior is a heavyweight metal die-cast figure. It comes with a portable launcher that shoots missiles. Acroyear, the enemy of these good Micronauts, is plastic and die-cast, with metal and plastic rollers for feet and wings that look like wheels.

Acroyear II is even more evil than his predecessor. He's made of die-cast metal. He has magnetic ball-joint fists, a detachable "flying pack," a snap-on missile-firing system, and two different sets of arms and legs. He can be changed into a missile-firing space vehicle.

Pharoid has detachable wings on his feet and a signal beacon on his breastplate that glows in the dark. He comes with his own time chamber.

Battery-powered Biotron is a robot and a vehicle combined. When the toy is a robot, it is 12" high. He can walk or roll on Caterpillar-style wheels. There is a special capsule in Biotron's chest that enables any Micronaut to attach itself to the robot and be carried. Biotron's legs can be detached and wheels added to turn him

OBERON

BARON KARZA

ANDROMEDA

into a space vehicle. His silver hands can grip objects with their spring-operated pincers.

Force Commander and his horse Oberon have "Magnopower" parts—built-in ball-joint magnets. They are white plastic and die-cast. Baron Karza, the Master of Evil, and his horse Andromeda, the Star Stallion, are all black. Baron Karza can fire missiles from his fist and chest. He

MEGAS

REPTO

comes with a detachable rocket launcher on his back. Andromeda has its own missile-firing system. Both are equipped with Magnopower parts, as well. Megas is another Magnopowered horse. He comes with a missile

launcher, and his legs can be exchanged for wheels to turn him into a rolling space vehicle.

Repto, Membros and Antron are part of the Micronauts Alien Invaders. All have glow-

in-the-dark brains. Repto, a reptilian android, has a plug-in laser saw and blaster and removable bat wings. Membros is a slithery-skinned android. He is equipped with a blaster and a plug-in backpack with

MEMBROS

ANTRON

GIANT ACROYEAR

HORNETROID

extension pipe for his Hydralaser. Antron is a six-limbed insectlike android. He has four tools and weapons that plug into his arm sockets: hook, eyebolt, neutralizer and laser blaster.

Giant Acroyear is King of all the Acroyears. He's 8½" high. His arms are two robots. His head and chest are a land cruiser with a detachable space glider. His legs combine into a swept-wing fighter plane. Hornetroid, over 19" long, is half insect, half mechanical vehi-

cle, known as Alien Invader. A Micronaut figure can sit behind its transparent eye case. Hornetroid shoots missiles from its tail, has spring-loaded jaws, and can flap its transparent wings by pressing a button.

Promotional materials also stated that Mego felt the future of Micronauts was "virtually limitless" and that new merchandise was being readied for following years. Mego hoped that Micronauts would become a staple for them and for retailers. But be-

cause of a changing market, the Kenner Star Wars juggernaut, and other factors, Mego went out of business in 1982 and many plans for the Micronauts were never realized. After the company's demise, Takara's Micro Man continued to be imported into the United States in similar packages under the name The Inter-Changeables. Micronauts will be reintroduced in 2000 under the direction of the Mego product developer, Martin Adams.

MIGHTY MORPHIN POWER RANGERS

BAN DAI, 1993–98
SIZE: 6" AND 9"
VALUE RANGE: $7–$35

JASON/RED RANGER
ZACH/BLACK RANGER
BILLY/BLUE RANGER
KIMBERLY/PINK RANGER
TRINI/YELLOW RANGER
SQUATT
GOLDAR
BABOO
KING SPHINX
FINSTER
PUTTY PATROL
MEGAZORD
DRAGONZORD
RHINO BLASTER
PIRANTIS HEAD
LORD ZEDD
WHITE TIGERZORD
TOR THE SHUTTLE ZORD
RED DRAGON THUNDERZORD
MOVIE EDITION PRODUCTS:
  HORNITOR
  IVAN OOZE
  SCORPITAN
  MEGAFALCONZORD
  FALCONZORD
  NINJA MEGAZORD
  NINJOR
  MERRICK THE BARBARIC
CALCIFIRE
MASTER VILE
SILENT KNIGHT
TALKING POWER RANGERS
DELUXE NINJA WARRIORS:
  DELUXE SHOGUN MEGAZORD
  DELUXE FALCONZORD
SERPENTERA
POWER RANGER ZEO
DELUXE ZEO MEGAZORD
DELUXE RED BATTLEZORD
DELUXE PYRAMIDAS

JASON/RED RANGER

ZACH/BLACK RANGER

The **Mighty Morphin Power Rangers** had been an average television show in Japan for several years before Saban Entertainment decided to try an Americanized version of the show. No one, except perhaps Saban, predicted that American kids would go bonkers for this live-action science-fiction adventure show. The American cast appealed to boys and girls, and the Power Rangers toys became the hottest-selling action figure line anyone had seen in years. For collectors, the 1993 series is the most valuable.

BILLY/BLUE RANGER

KIMBERLY/PINK RANGER

TRINI/YELLOW RANGER

In 1993, toy stores didn't have enough of the product. Many children went without their Power Rangers and Megazords for Christmas. Toy manu-

facturers rushed to Asia in search of look-alikes, and knockoffs appeared in toy stores as early as January 1994. But kids wanted the real thing.

In response to the demand, and embarrassed by the shortages, Ban Dai manufactured the line 24 hours a day in three different loca-

SQUATT

GOLDAR

BABOO

KING SPHINX

FINSTER

PUTTY PATROL

tions for several months. By the fall of 1994, supplies surpassed demand. In 1995, the Power Ranger movie helped sales. But Ban Dai realized that American kids wanted something different each season and so continued to introduce new characters and accessories each year.

Though the availability of the figures was limited, the first American year of the Power Rangers was a year of wild television success. The five original Power Rangers and various bad guys were the first of the line in stores.

The five Power Rangers were all 8" high. They all came in novel triangle-shaped window boxes with the Power Rangers logo above the window. Five photographs from the television show ran across the bottom, showing each character's human face transforming into its Power Ranger identity and further into the identity of the animal whose power the

MEGAZORD

DRAGONZORD

RHINO BLASTER

PIRANTIS HEAD

LORD ZEDD

WHITE TIGERZORD

TOR THE SHUTTLE ZORD

Ranger possesses. All were dressed in their Power Ranger jump-suits, with removable helmets so they can "morph."

The Red Ranger figure, Jason, has the power of a Tyrannosaurus Rex. He is a 17-year-old with a black belt in karate. Zach, the Black Ranger, has the power of the Mammoth. He's a musician and a dancer and an accomplished student of the martial arts. Billy, the Blue Ranger, has the power of the Triceratops. He is a brilliant teenager and is obsessed with knowing how every-thing works. Kimberly, the Pink Ranger, has the power of the Pterodactyl. She is a gymnast. Trini, the Yel-low Ranger, has the power of the Saber-tooth Tiger. She has "lightning hands and a peaceful soul," and the ability to neutral-ize any opponent with minimum effort.

In 1993, the line also included various evil space aliens, including Squatt, Goldar, Baboo, King Sphinx, Finster and Putty

Patrol. They were all 8" high and fully artic-ulated and were pack-aged in window boxes with the Power Rangers logo above the window and the Evil Space Alien logo below the window. In a circle to the lower right of the figure was the character's name. These aliens were col-orful and imaginative, and each came with an action feature.

The first season of the Power Rangers intro-duced the Zords—morphing robots that transported the Power Rangers and battled

RED DRAGON THUNDERZORD

the evil aliens. The Megazord is a large robot that morphs into five Dinozords that the Power Rangers can ride. Each Dinozord is the color of one of the Power Rangers. When the

HORNITOR

IVAN OOZE

SCORPITAN

MEGAFALCONZORD

Megazord and the Dragonzord come bubble-packed on color cards.

The toys were much more plentiful and varied in 1994. Ban Dai was now manufacturing bubble-packed 5½" figures, so the 8" figures were called "deluxe." They had more intricate window-box packaging, with a full-color picture of an alien. They are more intricately sculpted, and are fully jointed for poseability. They each

Megazord is assembled into one unit, it can fire power-punch missiles. The Dragonzord comes with the Green Ranger, who has power over it. It is a robotic dragon with a "power laser drill." Both the

FALCONZORD

NINJA MEGAZORD

NINJOR

MERRICK THE BARBARIC

come with a "unique alien weapon." Some of the new Deluxe Evil Space Aliens included Rhino Blaster, Pirantis Head, and Lord Zedd. Lord Zedd was introduced as the powerful archenemy of the

Power Rangers and leader of the evil space aliens.

The White Tigerzord and Tor the Shuttle Zord, part of the "power zord system," appeared in 1994.

They are heroic robots that have electronic light and sound capability. The White Tigerzord has a light-up mouth and makes tiger sounds. It morphs from a white robotic tiger into a

CALCIFIRE

MASTER VILE

SILENT KNIGHT

TALKING POWER RANGERS

DELUXE SHOGUN MEGAZORD

DELUXE FALCONZORD

thunder warrior. The included White Ranger figure can ride the Tigerzord. A white tiger collectible coin is also included in the package. Tor, the Shuttle Zord, is a larger Zord that can transport other power zords. Its dome lights up, and it makes electronic sounds. It can morph into a thunder warrior. Both of these Power Zords come in solid cardboard boxes, with a photograph of the toy on the front.

With the 1995 release of the **Mighty Morphin Power Rangers Movie**, a whole line of movie tie-in toys was introduced. Hornitor, Ivan Ooze and Scorpitan, three of the Evil Space Aliens featured in the film, were re-created in deluxe 8" action figures. Hornitor and Scorpitan had metallic, insectoid exoskeletons, and Ivan Ooze was sculpted in all his humanoid robed wickedness. All three are highly detailed and jointed for poseability. They come window-boxed, with the Power Rangers movie logo above the window and the name of the character below the window.

The Megafalconzord and the Falconzord figures from the movie are intricately designed. The Megafalconzord is a "ninja" Zord that morphs into a vehicle or a heroic robot. It comes window-boxed. The

SERPENTERA

Falconzord is a beautiful hawk-shaped robot that carries the White Ranger. It comes in a box with a photo of the toy on the front. Both have the movie edition logo and come with a collectible coin. The Deluxe Ninja Megazord is also a movie tie-in. All five Ninjazords are included with this figure.

Deluxe Auto Morphin Ninjor is a mysterious and powerful ally of the Power Rangers. He bestows the Power Rangers with more Zords, and teaches them the skills of the Ninjetti. This action figure is 10½" high and can morph from Ninja mode to battle-ready Ninja Attack mode. He comes with a Ninja sword. All the Auto Morphin figures have

POWER RANGER ZEO

DELUXE ZEO MEGAZORD

DELUXE RED BATTLEZORD

DELUXE PYRAMIDAS

push-button morphing action.

With the introduction of the Ninja concept to the line came newer 8" deluxe versions of some of the more popular Evil Space Aliens. Merrick

the Barbaric, Calcifire, Master Vile and Silent Knight were all introduced in 1995. All are window-boxed.

The year 1995 also saw the introduction of five talking Power Rangers. These figures were 8" high, window-boxed. Each figure speaks three different Power Ranger commands from the television show.

The Deluxe Ninja Warriors are Ninja Megazord, Shogun Megazord and Falcon-zord. These are the Zords that Ninjor created to replace the destroyed Thunder-zords. All of these figures have inter-changeable parts that can be combined in different ways to create different Ultra Zords. Separately, they are large robot figures that can morph into smaller Ninja-zords. Serpentera is

Lord Zedd's evil Zord that can morph from Dragon Mode to Attack Mode.

Power Ranger Zeo appeared in 1996. The Power Ranger Zeo line is essentially an English-language repackaging of the Japanese **Oh Ranger** line. Each of the six 5½" figures has a different fighting action and a weapon. The figures are blister-packed on color cards with the Power Rangers logo in gold and white. The Zeo Megazord and the Red Battlezord are also 5" high and similarly packaged. Pyramidas, the carrier Zord, can morph into two separate Zords.

A similar line using the banner of Power Rangers Turbo was produced in 1997, while Power Rangers Space was the 1998 marketing theme.

MINI MONSTERS

REMCO, 1983
SIZE: 3¾"
VALUE RANGE: $10 – $65

CREATURE FROM THE BLACK LAGOON

DRACULA

FRANKENSTEIN

MUMMY

PHANTOM OF THE OPERA

WOLFMAN

These 3¾" figures from Remco were based on the classic Universal Studios movie monsters. The original figures were plain, but later had glow-in-the-dark features. Each figure came packaged with a monster iron-on patch and a glow-in-the-dark skull ring. Each is nicely detailed and re-alistically painted, no mean feat for such a tiny figure. All have "monster-crush" grabbing action and are fully poseable and blister-packed. The color cards feature large photographs of

CREATURE FROM THE BLACK LAGOON

DRACULA

FRANKENSTEIN

MUMMY

PHANTOM OF THE OPERA

WOLFMAN

the original movie monster. Wolfman and the Mummy are the

hardest to find, but all figures in this series can be valuable. Accessories for this line

included a mini monster play case and a Monsterizer.

MOONRAKER

MEGO, 1979
SIZE: 12"

JAMES BOND
JAWS

These 12" figures of James Bond and his nemesis Jaws were manufactured by Mego as a tie-in to the very successful Bond film **Moonraker.** The James Bond figure is fully articulated and comes dressed in a sil-

JAMES BOND/$25–$85

JAWS/$175–$600

ver cloth space suit. Jaws, also fully articulated, is dressed in

street clothes. His main feature is his metallic mouth. Both figures are window-boxed. There is a color picture of Roger Moore as James Bond on the box, and the name of the character is right below the window.

Although the line is not particularly finely detailed, these figures are all quite valuable. Jaws is the most difficult to find because he was not sold in the U.S.

THE NOBLE KNIGHTS

MARX, 1968–72
SIZE: 12"
VALUE RANGE: $60–$140

SILVER KNIGHT
GOLD KNIGHT

The Noble Knights was a very high-quality line from Marx. Each figure was more complex and finely detailed than most later action figures. Each knight came with a fully detailed suit of 15th-century-style armor. The armor consisted of approximately 18 pieces, which had to be buckled onto the figure. Each figure came with three different hel-

SILVER KNIGHT

GOLD KNIGHT

mets, various weapons, a shield and a banner. An instruction sheet was included identifying all of the armor pieces by name.

The Silver Knight and the Gold Knight are each packaged in a cardboard box with a picture of the knight in full regalia defending a castle. This line is an exceptionally fine example of the kind of work Marx did in the 1960s. The figures today are wonderful additions to any collector's shelf and have been reproduced from the original molds. The 1968–1972 series were the only releases that have the values mentioned above.

OFFICIAL WORLD FAMOUS SUPER MONSTERS!

AHI/REMCO, 1973–76
SIZE: 8"
VALUE RANGE: $30–$175

CREATURE FROM THE BLACK LAGOON
THE MUMMY
WOLFMAN
COUNT DRACULA
FRANKENSTEIN

These action figures were based on the classic Aurora model kits, which in turn had been based on classic movie monsters. All figures were blister-packed on color

CREATURE FROM THE BLACK LAGOON

THE MUMMY

cards. On the back of the Kresge-styled (forerunner of Kmart) cards is artwork from the original Aurora Glow-Kit boxes. Each

WOLFMAN

COUNT DRACULA

FRANKENSTEIN

character had various running versions throughout the series. All the 8" figures are fully poseable. The dressed figures have cloth costumes.

The Creature from the Black Lagoon is the most valuable of the line. This green, wide-waisted version has plastic arms and legs.

A second version has a smaller waist and rubber limbs. There were three different versions of the Mummy. All are wrapped in bloody bandages and wear eye patches. The figure with the right arm across the chest is the most valuable. Wolfman had three variations. The figure with the fierce expression,

and hairy feet and hands, is the most sought after by collectors. There were five Count Draculas. The European figure came window-boxed and is the most valuable. Of the five different Frankensteins, the one with the "Boris Karloff head" is most valuable.

OFFICIAL WORLD'S GREATEST SUPER-HEROES!

MEGO, 1972–78
SIZE: 8"
VALUE RANGE: $12–$1,400

SUPERMAN
BATMAN WITH
REMOVABLE COWL
BATMAN
ROBIN WITH
REMOVABLE MASK
ROBIN
AQUAMAN
CAPTAIN AMERICA
SPIDER-MAN
TARZAN
SHAZAM!
GREEN ARROW
RIDDLER

PENGUIN
JOKER
SUPERGIRL
BATGIRL
WONDER WOMAN
CATWOMAN
GREEN GOBLIN!
THE HULK
THE FALCON
THE LIZARD
IRON MAN
MR. FANTASTIC
INVISIBLE GIRL
HUMAN TORCH
THE THING
THOR
CONAN
ISIS
KID FLASH
AQUALAD
WONDERGIRL
SPEEDY

SUPERMAN

**BATMAN WITH
REMOVABLE COWL**

BATMAN

**ROBIN WITH
REMOVABLE MASK**

ROBIN

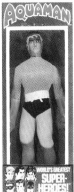

AQUAMAN **CAPTAIN AMERICA** **SPIDER-MAN** **TARZAN**

This long-running line of Mego action figures continued to sell well for the company until it went out of business in 1982. The year 1978 was the last one in which new figures and accessories were sold, but the toys were produced through 1982. Thirty-seven different super heroes, alter-ego identities and villains were included. A total of 33 characters were available boxed, and all

SHAZAM! **GREEN ARROW** **RIDDLER** **PENGUIN**

JOKER

SUPERGIRL

BATGIRL

WONDER WOMAN

but the alter-ego costumes were available on cards. There are numerous box, card, and figure variations, and hundreds of costume variations. All the figures are 8" high and dressed in cloth costumes. Most of the characters were from DC Comics, but there were also Marvel Comics and two independent characters in this series.

CATWOMAN

GREEN GOBLIN!

THE HULK

THE FALCON

THE LIZARD

IRON MAN

MR. FANTASTIC

INVISIBLE GIRL

Batman, Robin, Superman and Aquaman were introduced early in 1972. They were packaged in solid boxes. The color picture of the character on the back of the window box also appeared on the front of the solid boxes. The Batman figure with the removable cowl came in both box formats. Robin with removable mask has only been found in the solid box. Both versions appear

HUMAN TORCH

THE THING

THOR

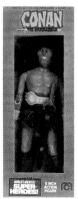

CONAN

ISIS

on Kresge-style cards. Tarzan and Captain America are advertised on solid boxes, but only window-box versions have been found.

Another way to recognize the earliest boxes in this series is the appearance of Captain America as the middle figure in the upper row of the box logo. He was replaced by Shazam on boxes for the first six DC characters in late 1972. The two Marvel figures were packaged in

boxes with pictures of Spider-Man and Captain America. Figures introduced in 1973 and later, have no character pictures on the front of the box, just the logo and wording. The series logo was changed in 1974.

Carded figures did not replace the boxed versions. From the first year of the series, the S.S. Kresge Company, forerunner to Kmart, demanded Peg-Board packaging. These "Kresge cards" are

KID FLASH

AQUALAD

216

WONDERGIRL

SPEEDY

worth 20% to 50% more than regular carded figures. Only Kresge-style cards were available in 1972, 1973 and early 1974. A version without the Kresge name was used by other retailers. Regular-style cards were introduced in 1974. The only figure not available boxed were the Teen Titans— Kid Flash, Aqualad, Wondergirl and Speedy. However,

boxed Isis figures are extremely rare.

In the 1970s, Batman, Spider-Man, Captain America and the Incredible Hulk were available in 12" figures, as stand-alone toys and with "fly-away action" (a piece of string and a sky hook to mount on furniture so the figure could be elevated from a remote location). A more muscular 12½" Spider-Man came carded with a web of netting.

A 9½" Robin and a Web-Shooting Spider-Man were also produced.

Although the Teen Titans are very rare, they were available through most retailers in 1976, the only year of their release. Vehicles and play sets were available for this line, made by Mego and other companies.

THE OUTER SPACE MEN

COLORFORMS, 1968–70
SIZE: 7"
VALUE RANGE: $125–$600

ALPHA 7
ELECTRON +
XODIAC
ORBITRON
COMMANDER COMET
ASTRO-NAUTILUS
COLOSSUS REX

In the late 1960s, everyone was jumping on the outer-space bandwagon, and Colorforms, famous for their plastic stick-on art sets for kids, was no exception. The Outer Space Men was one of the few three-dimensional toys produced by that company.

These outer space action figures were soft plastic "flexie" figures with accordion joints similar to Mattel's Major Matt Mason. The seven figures were from "the far-thest reaches of our galaxy." All came bubble-packed on a full-color card. The cards show photographs of the figures in action. Today, these figures are among the most sought-after collectibles by those who lived through the U.S. space race.

Each figure comes with a specialized weapon. Alpha 7, the Man from Mars, came with a ray gun. He is, of course, green, with a blue space suit. He is the smallest figure in the line.

ALPHA 7

ELECTRON +

XODIAC

ORBITRON

continued on page 221

218

Rare Outer Space Men

The Soviet Union shocked the free world in 1957 when they were the first to orbit the satellite **Sputnik**. It was the beginning of a great space race to put a man on the moon before 1970.

Against a background of rocket-booster testing, astronaut training, and the Mercury Space Program on the nightly news, Mattel developed Major MATT MASON, a "flexie" figure with black accordion molding at each joint of his space suit. Mattel's line was developed mainly to allow American kids to translate all the

real-life space adventures they were seeing on TV into play activity.

Colorforms saw an opportunity to offer a product that was very non-traditional for the company—a line of three-dimensional alien creature figures designed to complement the successful Major MATT MASON line.

The company asked freelance artist and toy designer Mel Birnkrant to create the line. His approach was to develop a creature for the other planets in our solar system with the exception of Mercury. Whereas Major MATT MASON was Earth's "Man in Space," Colorforms' goal was to provide representatives from neighboring planets. The packages were loaded with home-planet facts for added play value. Birnkrant used fictional accounts and scientific data as a basis for his

designs. The Mars representative was a little green man. The Man from Venus had wings like an angel.

The figures were a huge success. Customers clamored for more, and a second series of six figures, titled "The World of the Future" went all the way to the production phase. According to Birnkrant, only five sets were produced for the February Toy Fair of 1969. As a result of a New York dock strike, the figures got tied up in transit and were never shown.

On July 20, 1969, Neil Armstrong walked on the Moon, and the U.S. drive to become the space leader was fulfilled. Interest in space quickly faded, and the second series was scrapped. The figures shown here are one of two sets known to survive.

COMMANDER COMET

ASTRO-NAUTILUS

COLOSSUS REX

Electron +, the Man from Pluto, comes with a differently styled ray gun. He has a purple head and wears a silver space suit. Xodiac, the Man from Saturn, is packaged with a sword. He is red and wears a dark blue metallic suit.

Orbitron, the Man from Uranus, comes with a special detector. Commander Comet, the Man from Venus, comes with a crossbow. He is a winged figure with a human face. He wears a gold suit. Astro-Nautilus, the Man

from Neptune, carries a trident. He is all purple and has two legs and four arms. Colossus Rex, the Man from Jupiter, swings a mace. His reptilian skin is dark green and he wears a lighter green suit.

PEE-WEE'S PLAYHOUSE

MATCHBOX, 1988
SIZE: 6"
VALUE RANGE: $5–$15

PEE-WEE HERMAN

The whimsical and surreal live-action children's television series

Pee-Wee's Playhouse did more than jumpstart Laurence Fishburne's acting career, it saw the creation of a line of action figures. Matchbox sold figures based on the actors, and some of the "ani-

PEE-WEE HERMAN

mated inanimate" objects on the show. The fully poseable figure of Pee-Wee is dressed in his gray suit with red tie. He is bubble-packed on a full-color card, which echoes the Kenny Scharf set designs. Another Pee-Wee figure, with a helmet and scooter, was also available. (Note: Fishburne also played Cowboy Curtis.)

PLANET OF THE APES

MEGO, 1973–75
SIZE: 8"
VALUE RANGE: $30–$150

CORNELIUS
DR. ZAIUS
ZIRA
SOLDIER APE
ASTRONAUT
GALEN
GENERAL URSUS
GENERAL URKO
PETER BURKE
ALAN VERDON

CORNELIUS

DR. ZAIUS

ZIRA

SOLDIER APE

The classic science-fiction movie **Planet of the Apes** was one of the more influential science-fiction films of the late 1960s. It engendered several sequels and a television series, but the first, most would agree, is certainly the best. The Planet of the Apes action figures, based on the Charlton Heston character, are sought after by baby-boomer collectors. Few can forget their shock at the final scene.

ASTRONAUT

GALEN

GENERAL URSUS

GENERAL URKO

ASTRONAUT PETER BURKE

ALAN VERDON

The first five figures in this series, Cornelius, Dr. Zaius, Zira, Soldier Ape and Astronaut, appeared in 1973. The rest were produced in 1975. All the ape figures were produced in the 1970s, even though most items carry a 1967 copyright date. This is because the film characters were copyrighted that year.

These action figures are 8" high. The can be found bubble-packed or in window boxes. They are very nicely detailed, with cloth costumes faithful to the movie's. The Soldier Ape, General Ursus and General Urko are all armed with rifles and knives. The first-year figures are carded on a red card with the Planet of the Apes logo, and the second-year versions are carded on a yellow card. On the second-year cards, the astronaut helmet is replaced by the Mego logo, and the newer figures are featured.

PRINCESS OF POWER

MATTEL. 1985–87
SIZE: 6"
VALUE RANGE: $5–$15

SHE-RA
CATRA

Because He-Man was so successful for Mattel, the company wanted a spin-off and an expansion of this market. What better way to market to girls and compete with Galoob's Golden Girl series than to give He-Man a sister with her own action-figure line? Mattel introduced She-Ra in 1985. She is the "most powerful woman in the Universe," and is He-Man's twin sister. Her real name is Adora, and she transforms into She-Ra, the

SHE-RA

Princess of Power, with her own magic weapon, the Sword of Protection. Catra is one of her powerful and jealous rivals.

These figures are aimed at girls but attracted many male collectors. The female figures have "real" hair and flowing cloth cos-

CATRA

tumes. Many of the accessories are much more like a fashion doll's than an action figure's. She-Ra has "magic waist action," and Catra can be transformed into a cat. All the figures included a minicomic book.

RAMBO

COLECO, 1985–86
SIZE: 7"
VALUE RANGE: $6–$18

RAMBO
NOMAD

Rambo, the high-testosterone warrior introduced in the jingoistic Sylvester Stallone movie **First Blood,** was carried over to several bloody and violent sequels. The action figures, divided into two groups, "The Force of Freedom" and "S.A.V.A.G.E.," the enemies of Rambo, accurately capture the xenophobic spirit of the movies.

The Rambo figure from the first group, is heavily armed with several highly detailed weapons and a "battle action" rocket launcher. He looks a good deal like Stallone and wears the trademark headband. His

RAMBO

bare muscled torso is realistically sculpted to look like Stallone's famous body. He and his weapons are dual-bubble-packed on a card with both the Rambo and Force of Freedom logo.

Nomad is a figure from the S.A.V.A.G.E. series. In keeping with the antiforeigner tone of the movies, this figure reflects the worst stereotype of an Arab. Nomad is, according to the card, "devious, traitorous and desperate" and the "desert is his only home." He wears

NOMAD

desert fatigues, a white shirt, and an Arab headdress. He is given an Arabic-looking face with a mustache. He is heavily armed with battle action weapons. He is also bubble-packed on a card with the Rambo and S.A.V.A.G.E. logos. Both figures are packaged with a statistics card. The Nomad figure was discontinued due to Arab League protests in the media. The move resulted in greater popularity and value among collectors.

THE REAL GHOSTBUSTERS

KENNER, 1986–90
SIZE: 5"
VALUE RANGE: $5–$20
(HUMANS),
$10–$30 (GHOSTS)

PETER VENKMAN
RAY STANTZ
EGON SPENGLER
WINSTON ZEDDMORE
STAY-PUFT
MARSHMALLOW MAN
GREEN GHOST
MAIL FRAUD
TERROR TRASH
TOMBSTONE TACKLE
X-COP
GRANNY GROSS

PETER VENKMAN

RAY STANTZ

EGON SPENGLER

WINSTON ZEDDMORE

The 1984 blockbuster movie **Ghostbusters** appealed to both kids and their parents. At the time, it was the highest-grossing comedy film ever. The slapstick, cartoonish nature of the movie made it a natural for an animated series that picked up where the movie ended. Carrying over the main characters and ghosts, as well as adding some new ones, made for a funny and appealing television show. The producers had to call the show **The Real Ghostbusters** because the rights to the name "Ghostbusters"

STAY-PUFT
MARSHMALLOW MAN

GREEN GHOST

MAIL FRAUD

TERROR TRASH

TOMBSTONE TACKLE

X-COP

GRANNY GROSS

for an animated series were owned by Filmation.

The action figures and accessories were in the stores in plenty of time for the show's premiere in the fall of 1986 on ABC. Sixty-

five syndicated episodes aired in the fall of 1987, as well as thirteen new network episodes. Kids had the opportunity to watch the show six times a week—plenty of time to create the demand for the toys.

The team of Ghostbusters was made up of Peter Venkman, Ray Stantz, Egon Spengler and Winston Zeddmore. Each figure came with an ectoplasm weapon and a companion ghost. Venkman has a Grabber Ghost, Stantz has a Wrapper Ghost, Spengler has a Gulper Ghost and Zeddmore has a Chomper Ghost. These companion ghosts may prove to be collectible in the future. They get lost or separated from the figures so easily that there are bound to be more figures than ghosts. Each one of the four Ghostbusters is bubble-packed on a full-color card, with a picture of the cartoon character, as well as the famous Ghostbusters logo.

The Stay-Puft Marshmallow Man is a favorite figure among fans of the movie. This figure is the largest of the ghost figures. The Green Ghost, also known as "Slimer" on the television series, comes with pieces of plastic food. Mail Fraud, Terror Trash, Tombstone Tackle, X-Cop and Granny Gross are part of the Haunted Humans line. These figures transform from humans into ghosts.

THE REN AND STIMPY SHOW

MATTEL, 1993
SIZE: 5"
VALUE RANGE: $10–$25

REN
STIMPY

John Kricalfusi's hit animated series debuted on Nickelodeon in the early 1990s. It appealed to the sick and twisted side of kids and adults. It soon had a cult following, much like the show **South Park**. The action figures were based on specific episodes of the

REN

STIMPY

show. Slap-Happy Ren Hoek came with "wacky-whirling action" and was based on the episode "Stimpy's Busy Day." Bump-A-Riffic Stimpy was also based on this episode, and "smash-tastic" action. Kids were able to twist the figures' waists and then make the characters spin. The figures are bubble-packed on full-color cards, with drawings from the episode on which they are based. There is also a small photograph of the figure in action. Distribution of these figures was very limited because major retailers, except KB Toys, largely overlooked this line.

ROBIN HOOD AND HIS MERRY MEN

MEGO, 1974
SIZE: 8"
VALUE RANGE: $30–$200

ROBIN HOOD
LITTLE JOHN
FRIAR TUCK
WILL SCARLET

The four figures in this very nice but hard-to-find Mego series are based on the legend of Robin Hood. These 8" action figures are fully articulated and dressed in greatly detailed cloth costumes. Robin Hood, sporting a goatee and mustache, wears his trademark belted-and-fringed green tunic with yellow tights and hat, looking a little like Peter Pan. He carries his scabbard at his waist, and bow and arrow over his shoul-

ROBIN HOOD

LITTLE JOHN

FRIAR TUCK

WILL SCARLET

der. Little John has a full beard. He wears a brown tunic and carries a knife on his belt. Friar Tuck, the easiest in this series to find, wears the brown robe and hood of a monk.

Will Scarlet, the hardest figure to locate, is clean shaven. Of course, he wears a red tunic. He has his knife on his belt, and his bow slung over his shoulder.

The figures are boxed in attractive window packages. The character's name is above the figure, and the series name is below.

ROBOCOP, THE SERIES

TOY ISLAND, 1994–95
SIZE: 4½"
VALUE RANGE: $3–$10

ROBOCOP

This line was based on the science-fiction television series, which in turn, was based upon the film.

The Robocop figure was 4½" high. He is silver and comes with an M-16 battle rifle. The figure is bubble-packed on a card with a color drawing of the "future of law enforcement."

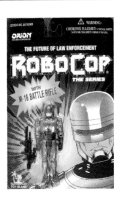

ROBOCOP

ROBOTECH

MATCHBOX, 1986;
HARMONY GOLD, 1992–94
SIZE: 3¾"
VALUE RANGE: $5–$35

RICK HUNTER
LISA HAYES
DANA STERLING
LUNK
MAX STERLING
ROOK BARTLEY

MIRIYA
ROBOTECH MASTER
BIOROID TERMINATOR
ZOR PRIME
MIRIYA

Three popular Japanese animated series, **Super-Dimension Fortress Macross, Super-Dimension**

Cavalry Southern Cross and **Genesis Climber Mospeada**, were combined and translated into American English to make the mid-1980s syndicated animated series **Robotech**. The producers originally wanted to bring only

RICK HUNTER

LISA HAYES

DANA STERLING

LUNK

MAX STERLING

ROOK BARTLEY

Fortress Macross to the United States, but there weren't enough episodes to create a syndication package, so the other two series were brought in. A "bridge" episode tied all three together into one narrative that dealt with three generations of the Robotech Wars. A Harris Poll taken in the mid-1990s showed that **Robotech** is one of the five best-remembered of all science-fiction shows.

The style of the cartoon is that of Japanese **anime**, and the action figures' packaging reflects that in Matchbox toys. All figures are blister-packed on a full-color card that features a

MIRIYA

ROBOTECH MASTER

BIOROID TERMINATOR

ZOR PRIME

MIRIYA

color picture of the television character and her or his accompanying robot. The Robotech logo is clearly visible with the Matchbox logo above it.

The Robotech Master and Zor Prime are the easiest to find. They are two of the enemies of the Robot Defense Force. Lunk, a member of the Robotech Defense Force, is the most difficult to locate. Miriya is "micronized" and

switches sides in the story, so two different figures of her were made. The Zentraedi Enemy version of Miriya is taller, and was shipped on a longer card than the 3¾" figures.

At the height of its popularity, there were **Robotech** fan clubs, comic books, novelizations and roleplaying games. The collectible action figures are an asset to any fan's collection.

SILVERHAWKS

KENNER, 1986–88
SIZE: 5"
VALUE RANGE: $5–$20

STARGAZER

BLUEGRASS

STEELHEART

QUICKSILVER

COPPER KIDD

STEELWILL

MOON STRYKER

CONDOR

WINDHAMMER

Kenner's SilverHawks was one of the first action-figure lines to widely use vacuum metalizing. The process, spraying molten metal onto the plastic form, became almost routine in later years.

Each SilverHawk figure has an action feature and comes with a companion "weapon bird," a weapon that transforms from a bird to a musical instrument, or from a bird into a weapon. Each toy is blister-packed on a silvery blue card, with the SilverHawks

STARGAZER

BLUEGRASS

STEELHEART

QUICKSILVER

logo and a drawing of the character.

Stargazer's left eye is a wide-angle lens. He is packaged with the weapon bird Sly-Bird, who comes with a magnifying glass. Blue-

grass has "bird launch action." He is packaged with Side Man, a guitar that transforms into a hawk. Steel-heart, a female figure, comes with metalized wings that snap out. She is packaged with

COPPER KIDD

STEELWILL

MOON STRYKER

CONDOR

WINDHAMMER

Rayzor, a weapon bird with tomahawk wing action. Quicksilver also comes with snap-out wings. His entire body is vac-metalized. He comes with Tally-Hawk, who has scissor-snap wings.

The Copper Kidd figure is a great example of the vac-metalizing process. His entire body is sprayed copper. He comes with snap-out wings and the weapon bird,

May-Day. May-Day comes with a whistle. Steelwill, another metalized figure, also has snap-out wings. His weapon bird is Stronghold, a hawk with vise-grip legs.

Moon Stryker comes with turbine waist and Tail-Spin, a weapon bird with rescue wings. Condor comes with a shooting left claw, and his weapon bird, Jet Stream, has double-wing eject. The larger figure Windhammer comes with Tuning Fork suitable for his size.

THE SIMPSONS

MATTEL, 1990
SIZE: 6"
VALUE RANGE: $8–$20

BART
HOMER
MARGE
LISA
MAGGIE
NELSON
BARTMAN

FOR KIDS OVER 4, MAN!

THE SIMPSONS
BART
WITH 5 COOL THINGS TO SAY.

AYE, CARUMBA!

MATTEL

BART

When historians examine 20th Century **fin de siècle** Western culture, they may come to the conclusion that an animated television show had a greater influence on popular culture than many "higher brow" works of art. In mid-1998, **Time** magazine named Bart Simpson one of the twenty most important cultural icons of the twentieth century, even over Mickey Mouse. It stands to reason that the Simpsons action figures will become even more important to

collectors as the years wear on.

The arch humor of the characters was captured nicely by Mattel in 1990, the second year of the Fox television series. The color-

ful action figures came with interchangeable plastic cartoon word balloons attached to the figures' heads. Each balloon had different phrases unique to the character. Each figure features a "per-

HOMER

MARGE

LISA

sonality" accessory: Bart has a skateboard, Homer comes with a radiation helmet, Marge has a sheet of cookies and an apron, Lisa comes with her saxophone, Maggie comes with her animal scooter, Nelson comes with a trash can, and Bartman, Bart's sometimes alter ego, comes with a cape and slingshot.

All figures are blister-packed on full-color cards with a picture of the character from the show and with the Simpsons logo on top. Additional figures for this series were planned, but never made.

MAGGIE

NELSON

BARTMAN

THE SIX MILLION DOLLAR MAN

KENNER, 1975–78
SIZE: 12"
VALUE RANGE: $30–$125

COLONEL STEVE AUSTIN
OSCAR GOLDMAN

Kenner created a line of action figures and accessories based on the very popular television series **The Six Million Dollar Man**, starring Lee Majors. The story was of an astronaut, Steve Austin, who was in a terrible accident. He is saved by scientists who replace his damaged eye, ear, arm and leg with "bionic" parts that give him superhuman strength, hearing and speed. Steve Austin becomes the first Bionic Man.

The Bionic Man figure is frequently found damaged or incomplete. The window-boxed figure included rubber skin on his right arm that could be rolled up to reveal plastic arm modules. The same arm could be removed and replaced with critical assignment arms. Oscar Goldman, the Bionic Man's boss, comes dressed in a classic early-1970s checked leisure-suit jacket and turtleneck. He carries an exploding briefcase that contains two file folders and a headset.

COLONEL STEVE AUSTIN

OSCAR GOLDMAN

SPAWN

TODD TOYS/McFARLANE
TOYS, 1994
SIZE: 6"
VALUE RANGE: $5–$50

SPAWN (CLAMSHELL
WITH COMIC)
MEDIEVAL SPAWN
CLOWN
ANGELA
PILOT SPAWN
MALEBOLGIA
COSMIC ANGELA
SPAWN II
VIOLATOR II
VERTEBREAKER
FUTURE SPAWN,
BLACK/RED
CY-GOR
THE MAXX
SHE-SPAWN
EXO-SKELETON SPAWN
ZOMBIE SPAWN
SPAWN III
MOVIE CLOWN
MANGA MEDIEVAL
SPAWN
MANGA VIOLATOR
MANGA ANGELA

SPAWN (CLAMSHELL
WITH COMIC)

MEDIEVAL SPAWN

CLOWN

ANGELA

To say that Todd Mc-
Farlane has revolution-
ized the action-figure
arena would not be
an overstatement. In
the short period since
Todd Toys introduced
its first **Spawn** action
figures at the 1994 Toy
Fair, his influence can
be seen throughout
the action figure
world. Not only are
the major toy compa-
nies showing more at-
tention to detail and
craft in their figures,
several smaller compa-
nies have emerged

PILOT SPAWN

MALEBOLGIA

COSMIC ANGELA

SPAWN II

VIOLATOR II

VERTEBREAKER

with "McFarlane-like" figures that really push the envelope. And the adult market for action figures has exploded.

The Spawn line was originally planned as a Mattel product. Dissatisfied with the company's plans for the line, creator Todd Mc-

Farlane formed his own toy company only nine weeks before the 1994 Toy Fair. Yet McFarlane was still able to get models to

FUTURE SPAWN, BLACK/RED

CY-GOR

THE MAXX

SHE-SPAWN

EXO-SKELETON SPAWN

ZOMBIE SPAWN

the fair. A favorite among collectors for its detail and faithfulness to McFarlane's **Spawn** comics, the company has been able to translate this artistic approach to other lines. Numerous color variations have been produced, including running changes and full repaints of the figures.

Early packages were marked with the company name "Todd Toys." This was changed to "McFarlane Toys" in 1995. The original first series figures were sold in a plastic clamshell with a **Spawn** comic book. Subsequent figures were packaged in the standard bubble-pack with a lightning bolt pattern or other themed backgrounds. The first six figures were repainted, and a paper insert covered the comic book to give it a more "family" look.

The live-action movie version of Spawn also "spawned" a line of figures, equal in quality and design to the comic-book–based figures. It is almost a dead certainty that all of the McFarlane figures will continue to be great additions to any collection.

SPAWN III

MOVIE CLOWN

MANGA MEDIEVAL SPAWN

MANGA VIOLATOR

MANGA ANGELA

SPIDER-MAN

TOY BIZ, 1994–96
SIZE: 5"
VALUE RANGE: $8–$28

**SPIDER-MAN,
WEB SHOOTER**

DR. OCTOPUS

CARNAGE

VENOM

PETER PARKER

KINGPIN

THE LIZARD

**10" SUPER POSEABLE
SPIDER-MAN**

GREEN GOBLIN

SHOCKER

SCORPION

THE RHINO

CAPTAIN AMERICA

ELECTRO

SPIDER-MAN, WEB SHOOTER

DR. OCTOPUS

CARNAGE

VENOM

Toy Biz broke Spider-Man out of the Marvel Super Heroes line in 1994. The first assortment of fully poseable figures was released in conjunction with a new animated series.

The Rhino figure with "head ramming action" was delayed and subsequently released in limited quantities, making it the most sought-after figure in the line. A number of Rhino figures on English/French cards were imported from Canada, but English-only packaging is preferred by most collectors.

Each colorful figure comes with a special action feature and is packaged on a full-color card with the Spider-Man logo and a color picture of the figure as it appears in the animated series. Each of these fun figures is well articulated and shows a good attention to detail.

Spider-Man, Web Shooter, came pack-

PETER PARKER

KINGPIN

THE LIZARD

**10″ SUPER POSEABLE
SPIDER-MAN**

four flexible gray tentacles attached to the figure's lower back. His muscular figure (much beefier than Marvel's Dr. Octopus) wears an orange-and-green suit and menacing dark sunglasses. Venom, the vigilante protector of innocents yet enemy of Spider-Man, has the same heavily muscled frame as Dr. Octopus, clad in a black suit with a white arachnid covering his chest. He features "jaw chomping action"—basically a hinged jaw with exposed fangs.

Carnage, one of Spider-Man's most dangerous enemies, was created by the spawn of the same symbiote that inhabited both Spider-Man and Venom. He wears a red-and-black suit with a hood that accentuates his fangs. In keeping with the comic-book charac-

aged in his traditional red-and-blue suit featuring a "web shooter" with a "web projectile" attached. Super Poseable Spider-Man features the traditional Spider-Man suit and has "super poseable action." His arms and wrists, as well as his

shoulder joints, waist and knees—unlike those of Spider-Man, Web Shooter—are flexible. Both versions come packaged with bonus collector pins.

Dr. Octopus's special feature is "tentacle whipping action"—

GREEN GOBLIN

SHOCKER

SCORPION

ter's ability to detach parts of his body and use them as weapons, the figure comes with "weapon arms and snap-on wrist accessories."

Peter Parker, Spider-Man's intelligent, unassuming human alter ego, wears a brown leather jacket and blue jeans and comes with a camera accessory, befitting his job as a reporter. Like Spider-Man, Web Shooter, he comes with a bonus collector pin.

Kingpin is literally the biggest bad guy of

Toy Biz's line. The massive figure of the cruel crime boss appears three times the size of the other figures. His huge body is dressed in a white sport coat, pink pocket square and tie, and blue pants. However, his head is the same size as those of the other figures in the line. His special feature is his "grab and smash action" arms with extra large hands.

The character of the Lizard first appeared over 30 years before in the Spider-Man comic-book series as a giant lizard-man

wreaking havoc on the innocent citizens of Florida's Everglades area. The Lizard's reptilian form features a removable lab coat over his purple pants and black torso. His action feature is his long, green "lashing tail."

The Green Goblin, Spider-Man's nefarious nemesis and one of the few who know of Spider-Man's secret identity as Peter Parker, is the most complex of the Toy Biz figures. He comes with the "goblin glider attack" feature (a blue plastic glider that

THE RHINO

CAPTAIN AMERICA

ELECTRO

attaches to his feet) and has "missile firing action," which allows him to shoot the enclosed pumpkin missile. His green body is dressed in purple boots, leotard, and matching skullcap and he carries a blue satchel.

Shocker's gold-and-red form is crisscrossed by a black lattice pattern that represents the vibroshock suit of the comic-book character. His action feature is "shooting power blasts"—he wears special silver-and-red gloves to which

enclosed pointed red missiles can attach.

Scorpion with "tail striking action" is a muscle-bound figure dressed in a green suit with a long, flexible tail. His tail, similar in construction to Dr. Octopus's tentacles, has a scorpion's stinger at its end.

The Captain America and Electro figures are packaged on cards different from those of the others in the series; the cards feature a picture of Electro-Spark Spider-Man. The

muscular figure of Captain America wears his traditional red, white and blue suit with hood, and comes with "real spark'n action" in the form of a sparkling shield and transforming hover jet. Electro comes with the same "real spark'n action" as Captain America. He is smaller in form, wears a green and gold suit with a gold-star–shaped helmet and comes packaged with a sparkling electro sled and arm cannon.

STAR TREK

MEGO, 1974–76, 1979 ($45-
$900); ERTL, 1984 ($5-$35);
GALOOB, 1988–89 ($8-$35);
PLAYMATES, 1992–96 ($4-
$15); ENESCO, 1994

SIZE: 3¾", 5", 8"

CAPT. KIRK
MR. SPOCK
NEPTUNIAN
ANDORIAN
THE ROMULAN
TALOS
ARCTURIAN
BETELGEUSIAN
MEGARITE
REGELLIAN
ZARANITE
LIEUTENANT TASHA YAR
(GALOOB)
CAPTAIN JEAN-LUC
PICARD
LIEUTENANT
COMMANDER DATA
DATHON
LIEUTENANT
COMMANDER
GEORDI LA FORGE
ROMULAN
VORGON
LIEUTENANT WORF
COUNSELOR
DEANNA TROI
AMBASSADOR SPOCK
CAPTAIN SCOTT
BENZITE
ESOQQ
Q IN JUDGE'S ROBE
CAPTAIN JEAN-LUC
PICARD (AS DIXON HILL)

AMBASSADOR K'EHLEYR
AMBASSADOR SAREK
BORG
THE HUNTER OF TOSK
QUARK
COMMANDER GUL DUKAT
ROM WITH NOG
KIRK
SULU
SPOCK
KHAN
DR. McCOY
LIEUTENANT UHURA
LIEUTENANT SAAVIK
KIRK IN SPACE SUIT
CHEKOV
GUINAN
ILIA PROBE
SWARM ALIEN

Probably no other se-
ries in entertainment
history has inspired
the sort of loyal,
dedicated following
among fans as has the
group of television

series and movies
known collectively
as **Star Trek**.

The original television
series, **Star Trek**,
brainchild of producer
Gene Roddenberry,
made its debut in Sep-
tember 1966. The 79
classic episodes in
that series have be-
come the most suc-
cessful in syndication,
shown in 94% of the
United States and in
over 75 countries. In
September 1987 the
Star Trek tradition
continued with the
debut of **Star Trek:
The Next Generation**,
which was the top-
rated drama series in
syndication for seven

CAPT. KIRK

MR. SPOCK

consecutive seasons, receiving 16 Emmy Awards as well as a Peabody Award for excellence in programming. In September 1993 **Star Trek: Deep Space Nine** reached television screens, followed in January 1995 by **Star Trek: Voyager**.

Aside from the four television series, seven **Star Trek** films have been made and have totaled revenues of over a half billion dollars at box offices worldwide. The more than 325 hours of Star Trek programming have generated merchandise sales of more than a billion dollars—stemming from novels, books on tape, videos, encyclopedias, electronics, clothing and other forms of merchandise. Conventions attended by Star Trek fans are held every weekend of every year. Even

NEPTUNIAN

ANDORIAN

THE ROMULAN

TALOS

NASA has been affected by the strength of Star Trek's popularity; the U.S. space shuttle **Enterprise** was named after the Star Trek flagship after NASA received 400,000 requests from fans known as "Trekkers."

Star Trek action figures were first introduced

by Mego in 1974, along with an Enterprise playset, and were followed by a series of four aliens in 1975 and an extremely rare second series of aliens in 1976. The figures measure 8" in height and feature removable clothing. This makes the figures, which are already much less stylized

ARCTURIAN

BETELGEUSIAN

MEGARITE

REGELLIAN

ZARANITE

than those of later manufacturers, appear more doll-like.

The series of six Mego figures begins with the character that made William Shatner

famous—Captain Kirk, Captain of the **Enterprise**. Kirk comes packaged on a multi-colored card featuring six circles containing pictures of each of the principal characters in

the original television series (Kirk, Mr. Spock, Dr. McCoy, Mr. Scott, Lt. Uhura, and the Klingon). He wears a yellow sweater with gold trim and the Star Trek symbol on the left shoulder. He has black pants and black molded boots and carries a phaser and communicator on his belt.

Mr. Spock, the supremely rational Vulcan science officer, comes packaged on the same card as Kirk. He wears the same style of Star Trek officer sweater, pants and boots as Kirk and is also outfitted with a phaser and communicator, but his sweater is blue. Unlike Kirk, Mr. Spock is packaged with a tricorder.

Neptunian, part of the 1975 Aliens series, comes packaged on the Star Trek Aliens card, which is printed with a red planet, blue moon, and a desert surface with mountains in the background. Neptunian is dressed in a green bodysuit decorated with "scales" and covered in a red leotard. He has green webbed feet and hands and a molded plastic head.

Andorian, The Romulan, and Talos, manufactured as part of the rare 1976 Aliens series, are packaged on a similar card to the 1975 Aliens figures. These figures are

LIEUTENANT TASHA YAR (GALOOB)

sought after by collectors and can fetch hundreds of dollars each.

Aside from the popular 8″ line of figures from the original **Star Trek** television series, Mego produced a series of 3¾″ figures from **Star Trek: The**

Motion Picture. The plastic figures are fully poseable and come on cards featuring the **Enterprise**, pictures of Kirk, Spock, McCoy, Decker, and Ilia, and the name of the character printed above the bubble-pack section.

Again, the aliens in this series were the most popular with collectors, including Arcturian, Betelgeusian, Megarite, Regellian and Zaranite. Of the figures in this series, only Betelgeusian and Megarite have nonplastic features— Betelgeusian a red cloth cape and Megarite a black hooded robe. Also, Betelgeusian's name does not appear above his figure on the card.

After the Mego Corporation collapsed, the licensing rights for Star Trek action figures passed to ERTL in 1984. ERTL manufactured only four figures from the motion picture **Star Trek III: The Search for Spock**. Of these, the only alien figure was a Klingon leader.

CAPTAIN JEAN-LUC PICARD

LIEUTENANT COMMANDER DATA

Galoob made Star Trek: The Next Generation products in 1988–89, with the Riker figure being the most common and the figures of Data and Tasha Yar the rarest—only two of each were enclosed in each carton of characters shipped to vendors. Some Data figures have color irregularities, resulting in figures with blue-green and spotted faces, which add to their value.

DATHON

LIEUTENANT COMMANDER GEORDI LA FORGE

ROMULAN

VORGON

The company's first line of figures from **Star Trek: The Next Generation** features ten characters from the series. Each poseable figure comes on a blue card featuring a picture of the

Enterprise and the featured character. The prized figures of Lieutenant Tasha Yar, Security Chief of the Starship **Enterprise**, and Lieutenant Commander Data, Android Officer, come with

phaser weapons and tricorder analyzers, as does the figure of Captain Jean-Luc Picard.

The Galoob line was a failure; largely because of the overproduction of the principal characters—the Riker figure was so common that toy stores were still trying to unload them several years later—and a scarcity of the popular alien characters.

With the advent of its line in 1992, Playmates began to produce a very diverse line of figures with superior sculpting and then branched out into sublines based on all the television series. Playmates has been the first company to understand the increased popularity of the alien figures among collectors and offers dozens. Previ-

LIEUTENANT WORF

COUNSELOR DEANNA TROI

AMBASSADOR SPOCK

CAPTAIN SCOTT

BENZITE

ESOQQ

cards, again featuring the **Enterprise**, the Star Trek: The Next Generation logo, and the slogan "Space. The Final Frontier." To further attract collectors, the cards advertise the fact that each figure in the series is individually numbered. All except for the first 11 figures in the series come with exclusive Skybox Playmates collector cards included in the packaging. Every figure in the series comes with an "Action Base."

The figure of Romulan, "the Imperialistic Enemy of the Federation," comes with three separate accessories—a phaser rifle, disruptor pistol and a Romulan Padd. He is dressed in a gray "quilted" Romulan-style jacket and black pants. The alien Vorgon, billed as "a Mysterious Alien Race

ous companies overproduced Kirks, Spocks, Jean-Lucs and Rikers while manufacturing shorter runs of the treasured alien characters. All characters are packaged with multiple accessories, and some of

the Playmates figures were also packaged with a "Space Cap" featuring the character, linking the figures to another popular form of collectibles.

All figures in these lines come packaged on brightly colored

From the Future," comes packaged on the same type of card, accompanied by "Vorgon Gear" consisting of tox uthat artifax, dilithium crystal, hex and a Vorgon scanner. His scaly form is brown with silver detailing, and his head is ridged with silver and pink markings. Dathon, Tamarian Captain, is of similarly "non-humanoid" background. He comes with a dagger, log book, flaming branch and Tamarian knife as accessories.

Lieutenant Geordi La Forge is dressed in the Next Generation's first-season uniform. The supersighted science officer is heavily accessorized and comes equipped with his trademark "V.I.S.O.R.," a metallic semicircular clip that gives the blind officer above-average-sight capabili-

Q IN JUDGE'S ROBE

CAPTAIN JEAN-LUC PICARD (AS DIXON HILL)

AMBASSADOR K'EHLEYR

AMBASSADOR SAREK

ties. In some figures the V.I.S.O.R. is removable instead of glued on—and more valuable. Lieutenant Worf, a Klingon Security Officer, wears the same style first-season uniform and is heavily accessorized with the

common phaser, tricorder and holster along with a ceremonial Bat'telh sword, Klingon combat blade, and Klingon sword. Unlike Geordi and Worf, the figure of Counselor Deanna Troi

BORG

THE HUNTER OF TOSK

QUARK

wears the second-season uniform.

Three of the characters from the original **Star Trek** series reappear in the Playmates Next Generation line, including Mr. Spock and Mr. Scott (now Captain Scott). Ambassador Spock is packaged with accessories from three different Star Trek races; he has a Klingon monitor, a Romulan Padd and phaser rifle, and a Vulcan book. Scottie, once Captain Kirk's steadfast Scottish Chief Engineer, ex-

hibits his Starfleet engineering roots through his accessories, including an engineering monitor and bio-engineering tools.

More specialized characters appear representing the different alien races, such as Mordoch the Benzite, "The First Starfleet Cadet from Benzar," and Esoqq, "A Member of the Chalnoth Race." Strikingly alien in appearance, they are indicative of the type of figures that help make the Play-

mates line so popular among collectors. Esoqq is especially valuable.

The enigmatic, omnipotent figure of Q is one of the most realistic in the series. This recurring character from the **Next Generation** series consistently challenges and frustrates Picard and the rest of the Starfleet officers. He appears dressed in the imposing red-and-black robes and red headdress of a judge, and his accessories are equally authorita-

COMMANDER GUL DUKAT

ROM WITH NOG

KIRK

tive—his "Continuum Gear" consists of a ceremonial gavel, lion statue, scroll and scepter. Star Trek figures were later done in 8" and 12" scales.

The episode that won the Peabody Award, "The Big Goodbye" is represented in the series by a figure of Picard dressed as Dixon Hill. The figure is basically the same as the standard Picard but dressed in a 1940s-style suit and bowler, and packaged with "1940's Style Accessories" such as a

pistol, rotary telephone, and floor lamp.

Ambassador K'Ehleyr, Klingon Ambassador to the Federation (and love interest of Lt. Worf), features the prominent ridged forehead of a female Klingon, and she wears lipstick. However, her accessories—a spiked glove, life-support mask, and ceremonial Klingon sword—are much less feminine. Ambassador Sarek appears on a style of card different from that of the earlier Play-

mates figures. The card is less colorful, the **Enterprise** featured on the front is more streamlined, and the slogan is omitted. The robes of the Vulcan Ambassador are notably detailed.

Playmates again changed the packaging for the Borg, eliminating the picture of the **Enterprise** in favor of a flashier Star Trek logo announcing the "Interstellar Action Series." The mechanical details of the Borg figure are well sculpted, and an action feature appears for the first time. The Borg comes with a "Spring Firing Cybernetic Arm" along with its other accessories. The Hunter of Tosk is packaged on the same card as the Borg, but not as a part of the "Interstellar Action Series." A Space Cap is included with the Hunter, but not with the Borg.

Among the more popular characters from the **Star Trek: Deep Space Nine** series are Quark and Commander Gul Dukat. They are packaged on purple cards featuring a picture of the space station with the Star Trek: Deep Space Nine logo and the slogan "Beyond the Final Frontier." Quark, the Ferengi Bar and Casino Proprietor, who taxes the patience of the characters on **Deep Space Nine**, comes

SULU

with five accessories, most notably a "Reptilian Pet" and an "Exotic Beverage Bottle." The valuable figure of Quark's Ferengi brother, Rom, with his Nog Mini-Action Figure included, came in two forms. The packaging and accessories of the two are identical with one exception: In one, the Nog figure is covered by the Star Trek: Deep Space Nine Space Cap, while the Nog figure in the other is visible. The more valuable of the two is the covered Rom with Nog.

SPOCK

KHAN

Playmates continued its line of figures with the production of the "Classic **Star Trek** Movie Series." The packaging consists of a colorful card with a fiery background and, like most of the other lines, a prominent picture of the **Enterprise**.

DR. McCOY

LIEUTENANT UHURA

LIEUTENANT SAAVIK

An "Exclusive Skybox-Playmates Deluxe **Star Trek** Movie Series Collectors Card" is included.

The characters Admiral Kirk, Lieutenant Sulu, Commander Spock, Dr. McCoy and Lieutenant Uhura are among the **Star Trek: The Motion Picture** figures featured in the line. Lieutenant Uhura is the most treasured as a collectible. Each well-sculpted figure is fully poseable and comes with "Galactic Accessories" particular to the character. Khan and Lieutenant Saavik are depicted as seen in **Star Trek II: The Wrath of Khan** and are packaged like the rest of the "Classic **Star Trek**" figures. Like Uhura, the female character, Lieutenant Saavik, has the most collectible value.

With the motion picture **Star Trek: Generations** Paramount Pictures and Viacom brought together the characters from the first two Star Trek television series—creating a much anticipated, successful film that had a larger gross in its opening weekend than any of the previous six Star Trek movies. The Star Trek: Generations series includes figures of Captain James T. Kirk in a space suit, Pavel A. Chekov and Guinan. The figures come on Star Trek: Generations cards with pictures of Captains Picard and Kirk, and Star Trek: Generations minimovie posters are included with each figure. The message "Hailing All Collectors" is printed on the card, reminding them that each figure in the series is individually numbered.

KIRK IN SPACE SUIT

CHEKOV

GUINAN

ILIA PROBE

SWARM ALIEN

The Chekov figure is by far the most valuable of the series. The former navigator of the original Starship **Enterprise** wears an updated version of his uniform and

comes with a "Classic" phaser, communicator, and tricorder. The figure of Kirk wearing a silver-and-black space suit is also very valuable to collectors, although less so than

Chekov. Guinan is a very good replica of Whoopi Goldberg as the Operator of the Ten Forward Lounge on the U.S.S. **Enterprise**. All three come with Star Trek: Generations gear.

The Ilia Probe and the Swarm Alien are part of the Serialized Warp Factor Series, and feature specialized Action Bases and accessories.

STAR WARS

KENNER, 1978–86
SIZE: 3¾"
VALUE RANGE: VARIES
DRAMATICALLY DUE TO
PACKAGE VARIATIONS

LUKE SKYWALKER

PRINCESS LEIA ORGANA

CHEWBACCA

ARTOO-DETOO (R2-D2)

HAN SOLO

SEE-THREEPIO (C-3PO)

STORMTROOPER

DARTH VADER

BEN KENOBI

JAWA

SANDPEOPLE

SNAGGLETOOTH

WALRUS MAN

BOBA FETT

LUKE SKYWALKER
(BESPIN FATIGUES)

IMPERIAL
STORMTROOPER
(HOTH GEAR)

LEIA (BESPIN)

IG-88

YODA

UGNAUGHT

LOBAT

HAN SOLO BESPIN

IMPERIAL TIE
FIGHTER PILOT

HAN SOLO IN
CARBONITE

EMPEROR'S
ROYAL GUARD

LUKE SKYWALKER
JEDI KNIGHT

LUKE SKYWALKER

PRINCESS LEIA ORGANA
(BOUSHH DISGUISE)

SQUID HEAD

ADMIRAL ACKBAR

TEEBO

ARTOO-DETOO WITH
POP-UP LIGHTSABER

ANAKIN SKYWALKER

LUKE SKYWALKER
IN IMPERIAL
STORMTROOPER OUTFIT

IMPERIAL GUNNER

A-WING PILOT

AMANAMAN

YAK FACE

SY SNOOTLES AND
THE REBO BAND

When the words "Long ago, in a galaxy far, far away" began to crawl across movie screens all over the United States on May 25, 1977, a phenomenon that would last for decades began, one which would change the toy industry forever. George Lucas's epic story, the first in a powerful and fantastic trilogy, created a new universe overwhelming in its scope and creativity—and extremely conducive to the manufacture of merchandise, particularly toys.

The action-packed, visually arresting film created a demand for Star Wars merchandise, but the toy companies were caught unprepared. Four months before the opening of the film, Bernie Loomis, then the president of Kenner, read the script and liked it enough to sign a contract. He had no idea how strong or how enduring the movie's popularity would prove to be, or how many figures, vehicles, and playsets would eventually be produced.

Even on a fast track, it takes a year to develop and produce a toy line. Therefore nothing could be ready for Christmas 1977 and Kenner was able to sell only a promise. For about $8, one could purchase a "early bird kit" featuring a certificate redeemable by mail for the first 4 figures in the series (Luke Skywalker, Princess Leia Organa of Alderan, Chewbacca, and Artoo-Detoo) and a cardboard "stage" that could accommodate the first 12 figures.

PRINCESS LEIA ORGANA

CHEWBACCA

ARTOO-DETOO (R2-D2)

It took Kenner over a year to catch up with enormous demand. Their number of figures would eventually climb to around 115 counting the TV Droids and other figures, but what really made the line special was the revolutionary change in the size of the figures. Previous action figures, like GI Joe and the Six Million Dollar Man, were 12" in height. The problem that toy executives faced with using this industry standard for Lucas's Star Wars characters and accessories was created by the scope and ambition of the movie itself—if the figures measured 12" in height, the vehicles and playsets necessary to accommodate them would be immense and prohibitively expensive for consumers. In a meeting with designers

HAN SOLO

SEE-THREEPIO (C-3PO)

STORMTROOPER

DARTH VADER

scheduled to discuss the logistics of production, Bernie Loomis is reported to have held up his right hand, thumb and forefinger apart, and said "How about that big?" at which point a de-

signer measured the distance at 3¾". Three and three-quarter inches became the height of Luke Skywalker, with other figures, playsets, and vehicles scaled to size. With the smaller figures selling at a lower cost of about $2 each, children were able to collect more of the figures.

The collectability of the line is very high, due in part to the

BEN KENOBI

amount of variation in the packaging and in the figures themselves—it is estimated that there are about 1,500 different figure/card combinations. Packaging and logo changes were made with the release of each of the films in the trilogy, and a number of variations in the figures were made over the years. The highly anticipated trilogy "pre-quel" released May 21, 1999 has launched another barrage of Star Wars figures, vehicles, and

playsets. (See Star Wars—Episode I.)

Luke Skywalker, the young hero of the series, has a number of figure variations, as well as the usual card changes. The blond Luke originally came with a telescoping yellow lightsaber, but after a few shipments the two lightsaber sections were remolded as one. Also, some of the Luke figures packaged on The Empire Strikes Back and Return of the Jedi cards have brown hair. The original blond Luke

JAWA

SANDPEOPLE

SNAGGLETOOTH

with the telescoping lightsaber is by far the most valuable. The Luke Skywalker figures with brown hair generally fetch more than those with blond hair, with the exception of the blond Luke packaged on the Star Wars card.

The Artoo-Detoo (R2-D2) figure was redesigned two times after the release of the original. Following the release of **The Empire Strikes Back**, the original R2-D2 figure came with a sensorscope. With **Return of the Jedi**, the android came with a pop-up lightsaber. Because of the figure variations, the R2-D2 with pop-up lightsaber is the most valuable. The last of the first 4 figures, Luke's sister Princess Leia Organa of Alderan (original version), is most valuable in the Return of the Jedi package, because by that time more Leia figures had been produced wearing different outfits.

The head of the macho Han Solo figure was first produced in a disproportionately small size. To correct this, a new Han mold was made, featuring a larger head. The original Han is available on Star Wars and The Empire Strikes Back cards and is quite rare, es-

WALRUS MAN

BOBA FETT

LUKE SKYWALKER (BESPIN FATIGUES)

IMPERIAL STORMTROOPER (HOTH GEAR)

pecially on the Empire card, and therefore very valuable. Versions of Han with the larger head came on The Empire Strikes Back cards, but are more common on a Return of the Jedi card with a photo variation.

The first See-Threepio (C-3PO) figure is a good likeness of the neurotic android. There are no figure variations until the Empire series, when C-3PO was retooled and began to feature removable limbs and a pack that fastened onto the back of other characters, allowing them to carry him. The original C-3PO remained available on the Empire card and is the most valuable of the C-3PO figures.

The Stormtrooper figure has remained the same throughout the series, and comes

LEIA (BESPIN)

IG-88

YODA

UGNAUGHT

with a black gun. Like most of the figures, Stormtroopers on the original Star Wars card are the most valuable.

The evolution of the figures of Ben Kenobi and Darth Vader, Jedi

Master foils who chose differently between the paths of good and evil, are similar to that of the Luke figure. All three original figures came with the telescoping lightsaber—Luke's yel-

LOBAT

HAN SOLO BESPIN

IMPERIAL TIE FIGHTER PILOT

HAN SOLO IN CARBONITE

stead of the later gray. The original figure is the most valuable of all 1,500 figure combinations in the entire series.

The original Jawa figure also commands high prices because of a major figure variation. The first Jawa figure produced is smaller than the later figure and dressed in a vinyl rather than a cloth cape. The tiny Jawa in the vinyl cape is worth over ten times the value of the figure in the cloth cape. Another interesting variation occurs in the "Sand People" figure. The name of the original figure "Sand People" was changed to "Sandpeople" with the Empire series, resulting in much higher prices for the "Sand People" label on the Star Wars card.

For Christmas 1978, Kenner made a Can-

low, Darth's red and Ben's blue—and all were modified at the same time to include a remolded single-piece lightsaber instead.

There are other variations in the Ben

Kenobi figure that make the original figure extraordinarily valuable. Aside from the telescoping lightsaber that was later modified, the original Ben Kenobi features white hair in-

EMPEROR'S ROYAL GUARD

LUKE SKYWALKER JEDI KNIGHT

tina Adventure Set, based on the popular scene from **Star Wars**, which was sold exclusively at Sears. A Snaggletooth figure was part of the set, but Kenner designers had only a photo of

the creature from the waist up to refer to when creating the character. They guessed wrong about the creature's stature and created a Snaggletooth that was taller and dressed in

royal blue with beige gloves and silver boots. The Snaggletooth later released on a card was much more true to the movie character—he is shorter, with a red suit and hairy hands and paws. The Walrus Man figure had no variations itself, but came with two slightly different-size weapons.

The reviled bounty hunter Boba Fett was originally offered as a mail-in promotional offer. The ads for the figure touted its spring-loaded missile-firing backpack. However, shortly before shipping, a wellspring of opposition was raised to a rocket-launching Battlestar Galactica toy that had reportedly injured some children. The Battlestar Galactica vehicle was recalled, and as a result, Boba Fett was changed. The

PRINCESS LEIA ORGANA (BOUSHH DISGUISE)

SQUID HEAD

already printed packaging promoting the backpack was covered up with solid black stickers. The figure arrived with the rocket permanently attached to the backpack and a note enclosed explaining the changes.

A figure of Luke Skywalker dressed in Bespin fatigues is included in The Empire Strikes Back line, the next major series after the original Star Wars line. The first Luke in Bespin fatigues had blond hair and came packaged on an Empire card with a white background. The card was subsequently reprinted with a different photo and background, and shortly afterward, as with the original Luke figure, the hair color was changed to sandy brown. Other figures from this part of the series include an Im-

ADMIRAL ACKBAR

TEEBO

perial Stormtrooper in Hoth Battle Gear, and Leia Organa in a "Bespin Gown." There are interesting variations in the packaging and manufacturing of the Leia figure—two cards were produced for

the Empire series, and the collar of her gown was produced in two colors, maroon and peach. The Leia figures on the card with the photo variation are the most valuable.

ARTOO-DETOO WITH POP-UP LIGHTSABER

ANAKIN SKYWALKER

LUKE SKYWALKER IN IMPERIAL STORMTROOPER OUTFIT

IMPERIAL GUNNER

A-WING PILOT

The IG-88 figure was initially labeled "IG-88: Bounty Hunter" in the Empire series, but in subsequent packaging the figure is referred to as simply IG-88. Conversely, the lovable Yoda is identified by name only in the Empire series, but titled "Yoda: Jedi Master" in the Jedi series. The original Yoda figure had an orange snake and cane, which was changed to a brown snake and cane. The Yoda with a brown snake and cane on the original

card is the most valuable.

The porcine Ugnaught figure is dressed in both blue-and-lavender smocks in the Empire and Jedi series. The Lobat and Han Solo Bespin figures remain consistent throughout the series, as does the Imperial TIE Fighter Pilot. The figure of the Emperor's Royal Guard appears only in the Jedi series.

Luke Skywalker was reissued in the Jedi series as "Luke Skywalker: Jedi Knight."

This Luke figure came dressed in the all-black outfit of a Jedi Knight, featuring a long brown cape and a snap-on gun and lightsaber. The figure was first shipped with a blue lightsaber, which was later changed to green. Other popular figures from the Jedi series include Princess Leia in Boushh disguise, Squid Head, Admiral Ackbar and Teebo.

After the Jedi series was completed in 1983, Kenner released a series of figures on new Power of the

AMANAMAN

YAK FACE

**SY SNOOTLES AND
THE REBO BAND**

Force cards. Each figure in the series includes a special collector's coin in the packaging. Many of the figures were Jedi characters exclusively released on new cards, and some were reissues of older figures. Han Solo in the Carbonite Chamber is part of the Power of the Force series and depicts Han deep-frozen in Carbonite.

A new Artoo-Detoo (R2-D2) with a pop-up lightsaber appears, as does Luke and Leia's father Anakin Skywalker, as he ap-

peared before he succumbed to the Dark Side of the Force and became Darth Vader. The figure of Anakin is one of the most valuable of the Power of the Force series, along with the figure of his son Luke dressed in an Imperial Stormtrooper uniform.

Most of the figures in the Power of the Force line are highly collectible, including the Imperial Gunner, A-Wing Pilot, Amanaman, and Yak Face figures. Yak Face was not commercially available in the United States—

the Yak Face figure with a weapon was available only in Kenner Canada packages and commands the highest price of all of the Power of the Force figures.

A 3-figure boxed set of Sy Snootles and the Rebo Band, who "really kept 'things' jumping in Jabba the Hutt's palace," comes with accessories like a microphone for lead singer Sy Snootles, a wind instrument for Droopy McCool, and an organ for Max Rebo.

269

The Blue Snaggletooth

The first **Star Wars** film released opened in the spring of 1977. Movie goers had never seen such special effects. The lines for weekend showings lasted for months as people returned to bring family and friends to such an amazing film.

Kenner had signed the deal to produce the **Star Wars** toys several months earlier, but the process of development was well under way before the film opened. George Lucas edited the film until very close to the time the negative was released to make prints. Without a print of the film for guidance, Kenner had to sculpt many figures using still photos as its only guide. This worked for all but one figure — Snaggletooth. The photo of Snaggletooth did not show the complete figure. This was particularly critical because the figure was to be included in a Sears' exclusive set. The deadline for the following Christmas catalog was April so the design had to be wrapped up before the film opened.

The toy designer assumed that the figure was full-sized although its legs were not pictured in the photograph. The designer tooled the Blue Snaggletooth figure for the set according to these assumptions. When the film was released it was eventually discovered this creature was midget-sized and the legs would have to be retooled. However, by the time of discovery Sears was committed to the picture in its catalog and production on the Blue Snaggletooth proceeded. Hundreds of thousands of these figures became available as part of the Sears' Exclusive Cantina Adventure Set. Until collectors got involved this inaccuracy went largely unnoticed because the scene in the film is so short.

The Red Snaggletooth wasn't scheduled for general release to another retailer until the third wave of figures. Therefore, time was available to make the necessary change. The correct version of the figure was the only one ever sold carded.

STAR WARS

KENNER, 1995–97
SIZE: 3¾"
VALUE RANGE FOR NON-
VARIATION FIGURES:
$5–$25 (RED CARDS),
$5–$18 (GREEN CARDS)

DARTH VADER

CHEWBACCA

BEN OBI-WAN KENOBI

C-3PO

BOBA FETT

LUKE JEDI KNIGHT

**HAN SOLO IN
CARBONITE BLOCK**

**LUKE IN STORMTROOPER
DISGUISE**

DEATH STAR GUNNER

CANTINA BAND MEMBER

EMPEROR PALPATINE

GRAND MOFF TARKIN

GARINDAN

GAMORREAN GUARD

YAK FACE

CEREMONIAL LUKE

**GENERAL LANDO
CALRISSIAN**

**DARTH VADER WITH
REMOVABLE HELMET**

BESPIN LUKE

**EWOK CEREMONIAL
PRINCESS LEIA**

R2-D2 WITH DATA LINK

Kenner began to mar-
ket new Star Wars fig-
ures in July 1995, using
the Power of the
Force slogan and fea-
turing new, fully re-
designed figures.

Darth Vader figures
with removable capes
came packaged on
the new Power of the

Force cards, which
had a picture of Darth
Vader's mask in the
upper left-hand cor-
ner and either a red or
green background—
the color of the card
was changed from
red/orange to green/
yellow in 1997. Again,
as with the original
Star Wars series, Ken-
ner initially had a difficult time with the
lightsabers for the
Darth Vader, Luke Sky-
walker and Ben (Obi-
Wan) Kenobi figures.
The first lightsabers
enclosed with figures
were too long, and in

DARTH VADER

CHEWBACCA

BEN OBI-WAN KENOBI

C-3PO

BOBA FETT

LUKE JEDI KNIGHT

1996 the lengths were reduced by approximately an inch. However, many different lengths appeared during the transition, until eventually the length was standardized. The

figures with the long lightsabers are the most valuable.

The Chewbacca figure comes with a bowcaster and a Heavy Blaster Rifle, and the

new C-3PO figure has a "Realistic Metalized Body," which reveals C-3PO's internal "wiring" at the waist.

The Ben (Obi-Wan) Kenobi figure, along

HAN SOLO IN CARBONITE BLOCK

LUKE IN STORMTROOPER DISGUISE

DEATH STAR GUNNER

with the lightsaber variation, has packaging variations. The original card had a typographic error on the "file card"—an informational card that could be cut out from the packaging listing the height, status, classification, affiliation, and weapon of choice of the figure, along with a couple of paragraphs of general information and the history of the character. When the typographic error was fixed, the photo on the back was changed, as well. The figure with the revised card back and the long lightsaber is the most valuable.

The Boba Fett figure was originally manufactured with black crescents on his gloves instead of circles—it was not a design change but a paint mask flaw that caused the variation. Thus, the Boba Fett figures that have crescents on the gloves are worth almost six times as much as those with the circles.

CANTINA BAND MEMBER

EMPEROR PALPATINE

GRAND MOFF TARKIN

GARINDAN

GAMORREAN GUARD

YAK FACE

CEREMONIAL LUKE

The Luke Jedi Knight figure was produced in three different forms. The new Luke figure, noticeably more muscular than the original Star Wars

Luke, came dressed in a Jedi Knight outfit, packaged with a lightsaber and a removable cloak. Some of the Luke figures have brown vests,

some black, and some of the black-vested Lukes were packaged in a special "Theater Edition" as part of a promotion surrounding the 20th anniver-

GENERAL LANDO CALRISSIAN

DARTH VADER WITH REMOVABLE HELMET

BESPIN LUKE

EWOK CEREMONIAL PRINCESS LEIA

R2-D2 WITH DATA LINK

sary release of the remastered **Star Wars: Special Edition** trilogy.

The figure of Han Solo frozen in Carbonite was packaged on two different cards. The figure and picture remain the same, but one card describes

the contents as "Han Solo in Carbonite with Carbon Freezing Chamber" and the other simply "Han Solo in Carbonite Block." The card upon which the figure is mounted also makes a difference in the collectible value of Luke Skywalker in Stormtrooper disguise and the Death Star Gunner; both are far more valuable on the

red card than on the later green one.

The Power of the Force series also included popular characters such as Emperor Palpatine, Grand Moff Tarkin, Garindan, Gamorrean Guard, Yak Face (Saelt-Marae), Ceremonial Luke, General Lando Calrissian, Bespin Luke and Ewok Ceremonial Princess Leia. Also produced at this time was the first Darth Vader figure to feature a removable helmet, as well as an R2-D2 with Data Link.

The Bith cantina band member was available as an exclusive mail-in offer from the Star Wars Fan Club.

SUPERMAN

MEGO, 1978–79
SIZE: 12"
VALUE RANGE: $35–$100

SUPERMAN

This handsome, cloth-outfitted figure was released with the first **Superman** film, along with three other figures. Mego planned to issue other figures but they were never produced, perhaps because many Superman figures were available as part of other lines.

SUPERMAN

SUPER POWERS

KENNER 1984–86
SIZE: 5"
VALUE RANGE: $10–$190

SUPERMAN
BATMAN
ROBIN
GREEN LANTERN
THE FLASH
AQUAMAN
HAWKMAN
WONDER WOMAN
THE JOKER
THE PENGUIN
GREEN ARROW
SHAZAM!
CYBORG
GOLDEN PHARAOH
MISTER MIRACLE
PLASTIC MAN

Kenner's Super Powers Collection of action figures is one of the most popular lines for collectors. Based on classic DC Super Hero action figures, sales of the line benefited greatly from its connection to a popular Saturday morning TV series. The Super Powers series followed the example set by Mattel's He-Man line—each figure was equipped with an action mechanism specific to the character. It was the first super hero line to do

SUPERMAN

so, a marketing approach that was echoed in the packaging. Each card depicts the figure's "power action" feature, making the line even more attractive to collectors. The line was initially

BATMAN

ROBIN

GREEN LANTERN

successful, but sales in 1986 fell dramatically after the introduction of both the Transformers and ThunderCats lines in 1985.

These figures are sculpted and accessorized very well. All are articulated at the head, arms, hips and knees. The Superman, Batman and Robin figures all come with capes and "power action" punches. Robin's punch is actually a karate chop.) Green Lantern, The Flash, Aquaman, Hawkman, and Wonder Woman round out the line of super heroes dedicated to fighting the residents of Akropolis—the bleak, desolate planet with a virtually impenetrable fortress, which is home to all of the villains.

The Joker and Penguin figures are remarkably true to the characters

THE FLASH

AQUAMAN

HAWKMAN

WONDER WOMAN

THE JOKER

THE PENGUIN

GREEN ARROW

SHAZAM!

CYBORG

as they appeared in comic books and cartoons. The tall, lean, laughing Joker and the short, stout, sneering Penguin are as dapper as they ever appeared in the comics, and their "power action madcap mallet" and "power action umbrella arm" features

are perfect accessories. The Green Arrow is an imposing figure, with a bow and arrow and, appropriately, a "power action archery pull" feature.

Shazam, Cyborg, Golden Pharaoh, Mr. Miracle and Plastic

Man were among new heroes released in 1986, along with an assortment of new villains. Cyborg "Steel Sentinel" is a bright metallic figure with "power action thrusting arms" that come with three attachments, and is the most valuable of the

GOLDEN PHARAOH

MISTER MIRACLE

PLASTIC MAN

line. The winged Golden Pharaoh, the chiseled Mr. Miracle, and the lanky Plastic Man figure, with a stretching neck, are

also prized by collectors.

Some of these figures were packaged on slim cards without the

illustration of the power-action feature and are much less valuable in this form.

SWAMP THING

KENNER, 1990–91
SIZE: 6"
VALUE RANGE: $15–$35

SWAMP THING

The DC comic hero Swamp Thing and other characters came alive in the animated TV series upon which this line was based. Swamp Thing was issued in four different

forms, including Snare Arm, Camouflage, Bio-Glow and Snap Up. The original Camouflage Swamp Thing had color-changing paint on both the arms and the chest, while the later version had only a painted chest. Each was released on a card that depicted the particular action feature of the "Thing."

SWAMP THING

TARZAN

MATTEL, 1978
SIZE: 9½"
VALUE RANGE: $50–$100

**TARZAN AND
THE GIANT APE**

**TARZAN AND
THE JUNGLE CAT**

R. DAKIN CO., 1984
SIZE: 4" AND 7"
VALUE RANGE: $10–$25

**TARZAN KING
OF THE APES**

TARZAN AND THE GIANT APE

TARZAN KING OF THE APES

TARZAN AND THE JUNGLE CAT

Mattel originally marketed these figures in window boxes labeled as "TV's famous lord of the jungle." Tarzan and the Giant Ape and Tarzan and the Jungle Cat came as two-figure sets featuring Tarzan, the animal, and a jungle knife. Tarzan and both animals were pick-ups from the 9" Big Jim line, repainted and stunningly repackaged under the famous Tarzan name. The Dakin Tarzan came in 7" and 4" figures made of jointed plastic, and Dakin also offered a 4" Young Tarzan figure made of bendable plastic.

TEENAGE MUTANT NINJA TURTLES

PLAYMATES, 1988–95
SIZE: 5"
VALUE RANGE: $3–$12
(EXCEPT "SCRATCH")

DONATELLO
LEONARDO
MICHAELANGELO
SPLINTER
APRIL O'NEIL
SHREDDER
ROCKSTEADY
KRANG
RAY FILLET
CHIEF LEO
CRAZY CLOWNIN' MIKE
DELTA TEAM DON
APRIL THE NINJA NEWSCASTER
DOCTOR EL
ROAD READY LEO
NIGHT NINJA RAPH
TOON MIKE
TURTLE TROLL RAPH
MIKE AS FRANKENSTEIN
DON AS DRACULA
RAPH AS THE MUMMY
LEO AS THE WOLFMAN
CARTWHEELIN' KARATE DON
SCRATCH
CAVE TURTLE LEO WITH DINGY DINO
KOWABUNGA CRACKIN' TURTLE EGG
STAR TREK CAPTAIN LEONARDO

STAR TREK FIRST OFFICER DONATELLO
APOLLO ASTRONAUT—LUNAR LEO
INVISIBLE MAN MIKE
THE MUTANT RAPH
BRIDE OF FRANKENSTEIN APRIL
CREATURE FROM THE BLACK LAGOON LEO

DONATELLO

The Teenage Mutant Ninja Turtles (TMNT) began as a comic-book spoof in 1984, and faster than you can say "Kowabunga, dude" they were a hit. Popularity among children grew as the pizza-loving "heroes in a half shell" named after Renaissance artists gained more exposure through

TV, film and merchandising.

The zany idea of these sewer-dwelling mutants held immense appeal for kids, and in 1988 the first ten figures in the line appeared, released on flashy, brightly colored cards featuring a prominent Turtles logo. Each of the four reptilian protagonists—Donatello, Leonardo, Michaelangelo and Raphael—came packaged with a trademark ninja weapon, as did their evil rat enemy Splinter. Some of the Turtles in the first run were manufactured with a soft (hollow) head—these are distinguishable by white fan club leaflets included inside and are worth up to three times as much as the hardheaded figures. The Turtles' friend April O'Neil was issued first in a yellow jumpsuit without the blue stripe on the sleeve, which was changed in the second run. April in the all-yellow suit is treasured by collectors. Shredder and Rocksteady are part of the first set of figures, as well.

LEONARDO

MICHAELANGELO

SPLINTER

APRIL O'NEIL

SHREDDER

ROCKSTEADY

KRANG

RAY FILLET

CHIEF LEO

CRAZY CLOWNIN' MIKE

this sewer-based fantasy is Krang—a large pink brain with mechanized "legs," a brain-drain gun and a mobile life-support unit included. The Ray Fillet figure, released later, advertised his "awesome mutant color change"—originally red and purple paint that was supposed to change to yellow and blue in cold water. However, the color-changing feature didn't always work as planned, since the colors faded rapidly and sometimes even changed in the package. Ray Fillet, "the fast fighting fish," was eventually just painted yellow and blue and packaged on a different card.

The Turtles line continued to expand rapidly to include figures like Chief Leo, which was part of the Wacky Wild West series, and

After the first ten figures were released, Playmates changed the packaging. In the next set, each figure came accessorized with its own "Wacky Weapon."

Typical of the outlandish characters in

DELTA TEAM DON

APRIL THE NINJA NEWSCASTER

DOCTOR EL

Crazy Clownin' Mike, which was part of the Bodacious Birthday series and featured a Michaelangelo figure dressed from head to foot in a brightly

colored clown costume. Further incarnations of the initial characters ranged from Delta Team Don, a camouflaged Donatello that was

part of the Mutant Military set, to April the Ninja Newscaster and Doctor El, the "plodding, peanut eatin' mini mammoth."

As popularity continued to stay strong over the years, more specialized Turtle sets were produced. In 1993, taking a page from Transformers, the Turtles line branched out into packaged sets like Road Ready Mutations including Road Ready Leo, a fireman Leo, who "mutated" into a fire truck, and Auto-Mutations including

ROAD READY LEO

NIGHT NINJA RAPH

TOON MIKE

Night Ninja Raph. Later a series of Toon Turtles was released, with characters like Toon Mike, who sticks out his tongue when a thumb wheel on his back is rotated. Turtle Troll Raph bears similarities to Toon Raph, but has a bright red head of troll hair. Cartwheelin' Karate Don is part of the Ninja Action series and can perform a midair cartwheel and land on his feet.

A Universal Studios Monsters series released in 1993 again mutated the turtles into different forms. Mike as Frankenstein, Don as Dracula, Raph as the Mummy, and Leo as the Wolfman are clever "turtlizations" of the famous movie monsters. A second set was released in 1995, featuring Invisible Man Mike, The Mutant Raph, Bride of Frankenstein April, and Creature from the Black Lagoon Leo.

TURTLE TROLL RAPH

MIKE AS FRANKENSTEIN

DON AS DRACULA

RAPH AS THE MUMMY

LEO AS THE WOLFMAN

CARTWHEELIN' KARATE DON

SCRATCH/$100–$300

The figure of Scratch was released with more regular series characters, also in 1993. But Playmates continued to produce far-fetched lines like Cave Turtles & Dinosaurs and Kowabunga Crackin' Turtle Eggs. Figures like Cave Turtle Leo and his Dingy Dino were marketed as "Stone Age collectible Turtle combos" and packaged in brightly colored window boxes. The Kowabunga eggs were fully motorized eggs that would blow apart and "hatch" a Turtle.

There seemed to be no end to Playmates'

CAVE TURTLE LEO WITH DINGY DINO

KOWABUNGA CRACKIN' TURTLE EGG

STAR TREK CAPTAIN LEONARDO

STAR TREK FIRST OFFICER DONATELLO

APOLLO ASTRONAUT— LUNAR LEO

INVISIBLE MAN MIKE

THE MUTANT RAPH

BRIDE OF FRANKENSTEIN APRIL

CREATURE FROM THE BLACK LAGOON LEO

creativity in marketing TMNT. The Turtles also appeared in Apollo 11 twenty-fifth anniversary packaging; like Lunar Leo, each turtle was dressed as an astronaut and included a TMNT collector card. Captain Leonardo and First Officer Donatello were part of 1994's Star Trek set. The Turtles were dressed in Star Trek officer outfits and included phasers, tricorders, communicators and Star Trek TMNT collector cards.

TERMINATOR 2

KENNER 1991–92, TOY
ISLAND, 1995
SIZE: 5½"
VALUE RANGE: $3–$22

**BATTLE DAMAGE
TERMINATOR**
**TECHNO-PUNCH
TERMINATOR**
BLASTER T-1000
JOHN CONNOR

On the heels of
Arnold Schwarzeneg-
ger's immensely popu-
lar movie sequel,
**Terminator 2: Judg-
ment Day,** Kenner
released a line of Ter-
minator figures. Based
on the violent, battle-
ravaged world of the
future depicted on
screen, these realistic-
looking figures mea-
sured 5½" tall and
were designed to
"demolish the compe-
tition." Kenner manu-
factured an initial line
of 11 figures, followed
by a second "Future
Wars" line of 8 figures.
Toy Island acquired
the rights in 1995 but

BATTLE DAMAGE TERMINATOR

TECHNO-PUNCH TERMINATOR

BLASTER T-1000

JOHN CONNOR

only produced 2
figures.

Many different forms
of the Terminator char-

acter appeared as fig-
ures, including Battle
Damage Terminator
with Blow-Open Chest
Action and Techno-

Punch Terminator with Super Smashing Action. The torso of Battle Damage Terminator "explodes" after a heavy hit to the chest, revealing the figure's cyborg skeleton. Techno-Punch Terminator has demonic red eyes and metallic body armor and throws a superpunch. The most powerful villain of the series is Blaster T-1000, which comes with rapid deploy missiles, or in a version featuring "Blast Apart Action" that causes the figure to blow apart into four pieces that can be reassembled "for even more battles." The John Connor figure comes with a motorcycle, allowing the young boy to accompany the Terminator in his journey to save humanity.

THUNDERCATS

LGN, 1985–87
SIZE: 8"

LION-O
MUMM-RA
CHEETARA WITH WILYKIT
TYGRA WITH WILYKAT
BEN GALI
JAGA
S-S-SLITHE
GRUNE

LION-O/$8–$65

MUMM-RA/$8–$65

ThunderCats was among the early product-based TV shows—mimicking the marketing and syndication strategy of **Masters of the Universe**. ThunderCats' production company offered television stations deals that guaranteed them a percentage of the profits of toy sales in their area. The show pitted the band of Thunder-Cats, led by Lion-O, against the evil forces of Mumm-Ra and his army of mutants. Mumm-Ra, when not in battle form, lurked in his black fortress as a withered, ancient mummy. The two sides were locked in an ongoing battle for

CHEETARA WITH WILYKIT/ $8–$65

TYGRA WITH WILYKAT/$8–$65

BEN GALI/$12–$35

JAGA/$12–$35

S-S-SLITHE/$8–$65

GRUNE/$8–$65

the powerful Sword of Omens.

All of the ThunderCats feature "Battle-Matic Action"—a "secret" lever on the back of the figure that makes the figure's arm move.

Cheetara and Tygra came with the Young ThunderCat figures of Wilykit and Wilykat. Some of the Tygra/ Wilykat sets featured a younger Tygra; however, they are equal in value to those that

come with the adult Tygra. Ben Gali and Jaga were also among the ThunderCats pride, while S-S-Slithe and Grune, The Destroyer, were evil mutant followers of Mumm-Ra.

THE TICK

BANDAI, 1995–96
SIZE: 7"
VALUE RANGE: $3–$10
($15–$55 FOR DIE
FLEDERMAUS AND "MAN
EATING" COW)

THE TICK

THE TICK

The Tick figures are based on the N.E.C. comic that spoofs comic books and has gained a following through an animated TV series that began in 1994. The "Bounding" Tick is an extremely muscular figure accompanied by amusingly named characters like "Pose Striking" Die Fledermaus, "Man Eating" Cow and "Grasping" El Seed, a matador with the head of a daisy. Series II offered two different versions of the Tick—"Hurling" and "Mucus." Die Fledermaus and the "Man Eating" Cow were produced in limited quantities and are five times as valuable as the other figures in the line.

TIM BURTON'S NIGHTMARE BEFORE CHRISTMAS

HASBRO, 1993
SIZE: 8½"

JACK
SALLY
MAYOR
OOGIE BOOGIE
LOCK, SHOCK AND BARREL

Tim Burton's Nightmare Before Christmas was a surprising hit at the box office, but the toy line, while prized by collectors, was not commercially successful. Owing to the short selling season, retail store buyers did not order nearly enough stock. The successful combination of Disney's renowned and unparalleled animation studios with the unique

JACK/$10–$60

SALLY/$10–$60

MAYOR/$10–$60

OOGIE BOOGIE/$50–$130

talent of filmmaker Tim Burton (who brought **Edward Scissorhands** to life) led to the creation of the whimsical, fiendishly funny world of Halloweentown. Its characters held an immense appeal for collectors, because of the characters' unique style and relation to the monster genre and because of the fact that retailers un-

LOCK, SHOCK AND BARREL/$35–$125

derestimated the film's potential when placing their orders. The movie was the surprise hit of the holiday season, creating a huge demand for the figures. To cut costs, the accessories later released with figures were packaged on plastic "trees" straight from the mold, but there is no difference in value.

The creepy yet somehow adorably appealing figures include accessories, and each has a special feature with instructions on the front of the card. In cartoonishly macabre detail, most come with a headstone as an accessory. The extremely tall, pencil-thin, bendable figure of Jack Skellington is dressed in a pinstriped tuxedo with a spider bow-tie, and has a pumpkin mask. Sally has detachable arms and legs, as well as a detachable head, and her limbs are covered with stitches. The nasty Oogie Boogie is a big fat monster with "creepy bug surprises inside," and the grinning figure of the Mayor cleverly features a "twist 'n turn two-faced head." Perhaps the most endearing, though, are Lock, Shock and Barrel—a trio of professional trick-or-treaters. Dressed as a witch, a devil, and a skeleton, these figures each have a mask and an accessory and come packaged as a set in a window box.

TOY STORY

THINKWAY, 1995–96
SIZE: 6"

FLYING BUZZ
QUICK DRAW WOODY
BO PEEP
GIANT BUZZ LIGHTYEAR

Toy Story, Disney's first wholly computer-animated feature film, was a huge hit at the box office in 1995. The figures of nursery toys that come to life and interact with each other are based around the characters of interloping astronaut Buzz Lightyear and cowboy Woody. Buzz Lightyear appeared in different manifestations includ-

FLYING BUZZ/$3–$7

QUICK DRAW WOODY/$3–$7

BO PEEP/$3–$7

GIANT BUZZ LIGHTYEAR/$10–$30

ing flying rocket action, karate chop action, and even in chrome. Woody was available with kicking or quick-draw action.

Bo Peep comes dressed in a cloth polka-dot dress with a bonnet and staff, accompanied by a three-headed sheep. The Giant Buzz Lightyear is packaged in one of two different window boxes: one with a logo strip across the plastic window, and a more valuable one with a yellow stripe. The Giant Buzz is a talking figure, whose box proclaims him to be the "ultimate talking action figure" and advertises his laser light feature along with his talking action.

TRANSFORMERS

HASBRO, 1983–96
SIZE: VARIES
VALUE RANGE:
$5–$70 (REGULAR CARDED),
$12–$90 (DELUXE CARDED),
$10–$50 (SMALL BOXES),
$15–$125 (LARGE BOXES),
$450–$700 (GIFT SETS)

BUMBLEBEE
CLIFFJUMPER
OPTIMUS PRIME
MEGATRON
STARSCREAM
THUNDERCRACKER
SKYWARP
DEVASTATOR
BLITZWING
ASTROTRAIN
WHIRL
SHOCKWAVE
GRIMLOCK
SLAG
SLUDGE
SWOOP
MENASOR
DEFENSOR
PIRANACON
SCOURGE
SKORPONOK
FORTRESS MAXIMUS
COUNTDOWN
INFERNO
JETFIRE
METROPLEX
TRYPTICON
SKY LYNX
COMPUTRON
OVERLORD

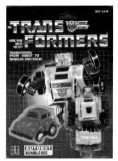

BUMBLEBEE

CLIFFJUMPER

"More than meets the eye" was the marketing cry for Transformers, a line of "robots in disguise" that became an instant success when introduced to American children in 1983.

Each robot was affiliated with one of two groups. The Autobots were heroic robots whose mission was to save planet Earth from the evil forces of the Decepticons. The two sides came from a war-scarred alien planet and continued to wage their battles on Earth. In the course of their conflict, the Transformers could change from their robot form into all kinds of mechanized vehicles, including sports cars, trucks, planes, boats, animals and weapons.

The line was a huge hit with children, who delighted in being able to change the robots back and forth from their robot to their disguise forms. Children's interest in the toys was encouraged by both a comic-book series and a popular television show, along with heavy print and TV advertising. In its sec-

OPTIMUS PRIME

MEGATRON

ond year, Hasbro had sales of over $300 million, but it was unable to sustain the growth over time, and in 1990 the first line was canceled in the U.S. only. Production never ceased for the international markets. Transformers released between 1983 and 1990 are known as Generation 1 Transformers and came in seven series. Most had a liquid crystal logo that proved the figure was a real Transformer.

Robots first released in 1983 were rereleased in 1984, each with a heat-sensitive rub sign that displayed the figure's allegiance. In 1986 the figures had many more metal parts, and all of the tires and wheels were made of either metal or rubber. Over time, to cut costs, many metal and rubber parts were replaced by plastic,

STARSCREAM

THUNDERCRACKER

SKYWARP

DEVASTATOR

resulting in a significant amount of variation in the figures.

Each figure comes with a bio card and a set of specs rating its strength, speed and skill, which can be deciphered using an enclosed decoder.

Bumblebee and Cliffjumper were among the first Transformers released—an assortment of minicars that transformed into robots. Their small, simplistic forms gave no indication of how complex and diverse

the line would grow to become.

The figure of Optimus Prime, the Autobot Commander, is a robot that transforms into a tractor trailer. The cab of the trailer changes into the robot figure of the leader, while

BLITZWING

ASTROTRAIN

the truck's trailer becomes the headquarters of the Autobots, complete with a scout car that races out of the back when a lever is pressed. This hefty figure is a playset and figure in one. Megatron, the Decepticon leader, comes in the form of a very realistic-looking Walther P-38 gun that transforms into a robot and a Particle Beam Cannon.

Starscream, Thundercracker and Skywarp are part of the Decepticon fleet of planes, designed by the Decepticons to fly in low, change into robot form, and attack the Autobots. The change from plane to robot is made by pulling back the wings and dropping the wheels, making it easy for children to switch back and forth.

WHIRL

SHOCKWAVE

Devastator is a clever six-figure assortment of Constructicons—robots that transformed into construction vehicles like cement mixers, dump trucks and bulldozers. Their function was to build Decepticon fortresses and energy plants.

The Constructicons were sold separately or in a gift set, and all six could be combined to form Devastator, one of a number of Super Warriors formed from multiple Transformers.

GRIMLOCK

SLAG

two. Their role was to confuse the Autobots with their three-way capabilities. Blitzwing changes from tank to robot to plane, while Astrotrain changes from space shuttle to robot to train. Whirl is another Autobot aircraft that transforms into a helicopter and comes with a "cage" cockpit, or a less valuable plain cockpit.

Shockwave transformed from a laser gun into a powerful robot, and sought to supplant Megatron as leader of the Decepticons. He came with "laser-action" lights and two "ultra laser" sounds requiring one 9-volt battery.

Grimlock, Slag, and Sludge were part of Series 2's Dinobots assortment—a set of five dinosaur Transformers that combined prehistoric figures and

The Triple Changers assortment of figures included Blitzwing and Astrotrain, which were designed to transform into three different forms instead of the usual

SLUDGE

SWOOP

futuristic design. They came packaged in their dinosaur form— Grimlock a Tyrannosaurus Rex, Slag a Triceratops and Sludge a Brontosaurus. The Pterodactyl Dinobot, Swoop, is the most valuable of the Dinobot assortment.

New additions in the next series included reflective patches and glow-in-the-dark posters that were packaged with some of the figures. Like Devastator, Series 3's Menasor was a Decepticon that could be created with the 5 smaller Stunticon figures. The Stunticons were based on cars like Porsches and Lamborghinis and were each sold separately in packaging with a picture of Menasor.

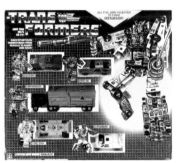

MENASOR

DEFENSOR

PIRANACON

SCOURGE

Defensor is another Super Warrior figure sold as a set. Defensor is actually a combination of the five Protectobots. Each Protectobot represents a law-enforcement vehicle, and all five band together to form the super-robot.

Metroplex is an Autobot robot that transforms into a stronghold with an attack car, a car launcher, and towers that transform into a robot and an attack tank. Trypticon is the Decepticon version, which transforms into a walking dinosaur.

Trypticon has all sorts of battery-operated action features like radar dishes, lighted sentry towers, flashing laser cannons and rotating gun turrets.

Sky Lynx is a space shuttle with a transporter. The transporter changes into a walking lynx, while the shuttle changes into a birdlike "swooping Autobot defender." Sky Lynx is actually motorized—the wheels of the transporter are battery powered.

Scourge is a Decepticon hovercraft popu-

SKORPONOK

FORTRESS MAXIMUS

AUTOBOT COUNTDOWN

COUNTDOWN

lar in the Targetmaster assortment—which, along with the Headmaster assortment, form much of Series four. Targetmasters are a group of Transformers based on aircraft. They come with

INFERNO

accessories that transform into mini-Nebulan figures. In Transformer lore, Nebulos was another planet to which a group of the Autobots fled, having tired of the incessant battling with the Decepticons. The planet had been peaceful for over 10,000 years, and its inhabitants feared and mistrusted the Autobots when they arrived. A group of Decepticons followed the Autobots to Nebulos and captured Lord Zarak, the Nebu-

lan leader. Both Autobots and Decepticons bio-engineered the Nebulans to be able to transform into weapons and Transformer heads, and the battle continued on the formerly peaceful planet.

JETFIRE

METROPLEX

Skorponok is part of the Headmaster assortment, and comes with two other Transformers included, Lord Zarak and Fasttrack. This powerful Decepticon transforms from a scorpion with movable legs and claws into a robot. The Lord Zarak figure can ride the scorpion in a special cockpit and transforms into the head of Skorponok when the figure is in its robot form.

The Fasttrack figure transforms into a tank and is included along with other defense bay features. Fortress Maximus is another in the Headmasters assortment, and the leader of the Nebulan-based Autobots. The figure of Fortress Maximus is two feet tall and transforms from robot to armored battle station to a defense fort, with all kinds of special features like a movable car elevator, push-button launchers, and a detention center. The Nebulan figure included with Fortress Maximus transforms into Cerebros, which becomes the head of Fortress Maximus.

Computron is similar to many of the other Super Warrior figures. The figure is made up of five specialized vehicles called Tech-

TRYPTICON

SKY LYNX

COMPUTRON

nobots that combine to form Computron.

Piranacon was released in 1988. The six-figure set of Seacons transformed from monstrous fish to robots to weapons, and combined to form the large Piranacon figure. The Micromaster assortment, with the Countdown command center/rocket launcher, was also released this year.

Inferno is part of the Action Masters assortment, which was released in 1989 as part of Series seven. Each of the Action Masters is a specialized Transformer with an action feature—Inferno carries a hydro-pack, which converts from a backpack into a water laser. This was the last year of Generation one in the United States; however, two more small series were produced and released in the

OVERLORD

continued on page 307

Tranformers: Robots in Disguise

Changeable robot toys had been popular in Japan since the mid-1970s prompting the production of an animated television series. In America, the FCC's rule, that a toy could star in its own TV series, sent toy executives scouring the world in search of the "star" toys to bring to America. A particular team of toy developers from Hasbro was flying high after the successful 1982 re-introduction of **G.I. Joe** in a 3¾" size. During the Tokyo Toy Fair that year, this elated team spotted several die-cast metal robots developed by Takara and Bandai at the Tokyo Toy Fair. The expressive tooling had been done, lots of animated TV programming was already produced, and both companies were eager to sell the rights for the vast U.S. market.

There were two major areas of concern. The first was the question of kid acceptability in the U.S. There had never been any toys like this before. Hindsight indicates there was nothing to be concerned about, but with the U.S. product launch estimated at $10 million, and your job on the line, you

act with caution. Secondly, these complicated toys had to be sold in 30 second commercials on television. **Transformers—Robots in Disguise** proved to be a winning catch phrase. Its animation made it look like the toys were changing from machine mode to robot mode all by themselves. When the actual toys were easy for the kids to "transform," but near impossible for the parents to work, Hasbro had a second mega-hit toy in as many years. Nowhere in modern toy history has a feat this big been equaled—before or since—the 1982/83 3¾" **G.I. Joe/Transformers** launchings. It's one of the key reasons Hasbro is one of the two giant U.S. toy companies at the beginning of a new century.

United Kingdom. These included Overlord, a massive Decepticon figure that included two Energons—a jet and a tank—and transformed into a playset.

Generation two Transformers were first released on revamped cards in 1993. Two years later Generation two featured a Cyberjets assortment including Jetfire, a gray camouflage F-15 fighter plane with a shooting missile launcher. Generation two was replaced in 1996 with the next incarnation of Transformers, the Beast Wars series.

BEAST WARS TRANSFORMERS

KENNER, 1996
SIZE: VARIES
VALUE RANGE: $3–$10

TERRORSAUR
DINOBOT
RATTRAP
INSECTICON
WASPINATOR
TARANTULAS
POLAR CLAW
OPTIMUS PRIMAL
B'BOOM

TERRORSAUR

DINOBOT

Beast Wars is a line of Robot warriors that are disguised as wild animals, insects and fish. Instead of Autobots and Decepticons, the heroic Maximals fight against the evil Predacons. The mechanized cards of the two generations of Transformers disappeared, and figures were packaged on new Beast Wars cards, featuring a new logo and a flashier design.

RATTRAP

INSECTICON

A variation on Generation one's Swoop, Terrorsaur was a pterodactyl who turned into an aerial

WASPINATOR

combat robot, with quick attack action, and Dinobot is a "velocoraptor."

Rattrap is a rat that transforms into a robot spy; Insecticon

TARANTULAS

a beetle that turns into a surveillance robot; Waspinator a wasp that features a shooting attack feature; and Tarantulas features a shooting

POLAR CLAW

OPTIMUS PRIMAL

B'BOOM

web launcher. These figures are nowhere as creative or as complicated as most of the Transformers and are designed with a much less mechanized look.

Mega Beast Polar Claw is a polar bear that fires a hidden robotic bat.

B'Boom is a mandrill that transforms into a

robot expert in guerrilla warfare, and Optimus Primal is a gorilla that turns into a robot that launches missile-blasting cannons from his arms.

ULTRAMAN

BANDAI, DBA
DREAMWORKS, 1991
SIZE: 8½"

ULTRAMAN

Popular in Japan, Ultraman is an action hero whose mission is to defend the universe and whose secret identity is Jack Shindo. Hundreds of action figures based on the Ultraman char-

acter have been marketed overseas. Seven figures were produced for marketing in the United States, but the line did not take off. The muscular, well-modeled figure of Ultraman came with a small figure of Jack Shindo in either a blue or a red outfit. A fighter jet and five "Ultraman intruder" figures rounded out the line.

ULTRAMAN

UNIVERSAL MONSTERS

REMCO, 1980
SIZE: 8"
VALUE RANGE: $50–$350

FRANKENSTEIN
CREATURE FROM THE BLACK LAGOON
MUMMY
WOLFMAN

The figures of famous movie monsters in this valuable series are fully poseable and come packaged in window boxes advertising their "Monster-Crush grabbing action." With the push of a button on its back, each figure reaches out to grab other figures. Parts of each figure glow in the dark, and they come

FRANKENSTEIN

with iron-on patches and glow-in-the-dark skull-and-crossbones rings. Each is dressed in cloth costumes. The Creature from the Black

Lagoon was produced later than the Frankenstein, Mummy and Wolfman figures, and is worth much more.

CREATURE FROM THE BLACK LAGOON

MUMMY

WOLFMAN

Universal Monsters: An Identity Crisis

The Universal Studios produced most of the famous monster films of the 1930s—**Frankenstein**, **Dracula**, **Wolfman**, **The Mummy**, **The Creature of the Black Lagoon** and several others. These characters, except for **The Creature of the Black Lagoon**, were achieved by using makeup. **The Creature** came to life by an actor donning a unique rubber suit. Each actor, however, did reveal his own image when they got into character. The make up and costumes did not fully hide who they were. Years later the ownership of the original characters raised many legal issues. When the original actors signed contracts to do the films there was no provision for licensing. By the time the market for licensing was realized the actors had all died.

The matter went unsolved for over 40 years. Whenever the character was reproduced as an action figure or mask, it resembled the original actor's image closely. The families sued, claiming that they had the rights, as heirs, to license the monsters. Of course, the more the monsters were produced and distributed, the more popular they became. The demand was driven high up, especially around Halloween time. In 1977 the matter

was resolved in an interesting compromise. The families permitted U.S. postal stamps to be printed with an exact likeness of the original actors. Since then Universal and the actors' families have jointly licensed the characters on one hand and the actor's image in the makeup on the other.

Some of the first action-figure products to be produced under this arrangement were made by Hasbro (12″) and Sideshow (7″).

VENOM, ALONG CAME A SPIDER

TOY BIZ, 1996
SIZE: 6"
VALUE RANGE: $4–$12

BRIDE OF VENOM
SPIDER-CARNAGE

In this series based upon Venom, the malevolent Spider-Man character, Bride of Venom and Spider-Carnage figures come on elaborate black, purple and green

BRIDE OF VENOM

SPIDER-CARNAGE

arachnid cards. Both figures are mostly black and well de-tailed and come with

Vile and Spit spider figures that have "snap attack legs."

WETWORKS

MCFARLANE TOYS, 1995–96
SIZE: 8"
VALUE RANGE: $5–$40

VAMPIRE
WEREWOLF
FRANKENSTEIN
BLOOD QUEEN

The third line of fig-ures introduced by McFarlane Toys, Whilce Portacio's Wet-works is a line of heav-ily accessorized,

powerfully built "Ultra Action" figures with specialized action fea-tures. Vampire is a frightening figure al-most 8" tall, with black hair, extended sharp claws and removable leg armor. Werewolf, a large red figure with a powerful, elongated torso, can stand on two legs or drop to all fours. Some Werewolf figures were painted

VAMPIRE

WEREWOLF

FRANKENSTEIN

BLOOD QUEEN

gray and are more valuable. Fewer Werewolf figures were produced than the rest of the line, making the

figure instantly valuable with collectors.

Blood Queen and Frankenstein are part of the second series

of Wetworks figures, which are packaged with lightning bolts on purple instead of blue cards. The entire second series of figures was repainted, and the repainted figures are actually more valuable than the originals. Blood Queen is a figure of a disproportionately tall woman, with spiked gloves and strong legs clad in thigh-high black boots. She comes packaged with a ceremonial staff and a cloth cape. The repainted figure is the same, but parts of her black outfit are trimmed in red. Frankenstein changes more drastically—the original brown, gray, and ivory figure is repainted in shades of green.

WHO FRAMED ROGER RABBIT?

LGN, 1988
SIZE: 6"

ROGER RABBIT
JESSICA

These figures based on the popular Disney movie **Who Framed Roger Rabbit?** came in two forms—the "Animates" were made from regular plastic, while the "Flexies" were made from bendable plastic. The

ROGER RABBIT/$3–$12

Roger figure comes with a set of handcuffs, and figures of his wife, Jessica, were

JESSICA/$25–$50

pulled by Disney because the figure was deemed "too sexy."

WILDC.A.T.S.

PLAYMATES, 1995–96
SIZE: 6"
VALUE RANGE: $6–$18

VOID
VOODOO

VOID

VOODOO

The WildC.A.T.S. line, based on a super-hero team from Image Comics that branched out into a Saturday morning TV show, sold well in its first year but was only carried at Wal-Mart in its second. The story line of the series is based on two groups of warring aliens that se-cretly exist within human society, searching for a powerful Orb. The Daemonites wish to use the Orb for Evil, while the Kherubim look to save the earth and the universe from evil. The female figures of Void and Voodoo are the most valuable in the line.

WIZARD OF OZ

MEGO, 1974; MULTI-TOYS, 1989
SIZE: 8"
VALUE RANGE: $20–$55

DOROTHY WITH TOTO
SCARECROW
TIN WOODSMAN
COWARDLY LION

In 1939 **The Wizard of Oz** film was released, surprisingly to mixed reviews, but its popularity has snowballed over the sixty years that have passed. CBS brought the film to TV screens annually through the 1970s and 1980s, and from 1974 to 1976 Mego produced a series of figures based on the beloved movie.

The Mego figures came packaged in green window boxes, and the figures wore real cloth outfits—except the Tin Woodsman, whose "tin" is really plastic. Dorothy is dressed in her

DOROTHY WITH TOTO **SCARECROW** **TIN WOODSMAN** **COWARDLY LION**

trademark white-and-blue checkered dress, the Scarecrow wears a scruffy black suit with "straw" accents, and the Cowardly Lion has a furry brown suit and a medal for courage.

In 1989, to celebrate the fiftieth anniversary of the film, Multi-Toys produced three lines of figures in different sizes—3¾", 6" and 12".

WONDER WOMAN

MEGO, 1978
SIZE: 12¼"
VALUE RANGE: $75–$175

WONDER WOMAN

This 12¼" Mego figure was produced during the run of the popular 1970s TV show **Wonder Woman**. The figure of Lynda Carter as Wonder Woman is packaged in a window box with a card attached to the side. The Wonder Woman figure comes dressed in her famous red, white, blue and gold leotard and red boots. The outfit of Diana Prince, her secret identity, is included.

WONDER WOMAN

316

WWF—WORLD WRESTLING FEDERATION

HASBRO, 1990–94
SIZE: 4⅞"

HULK HOGAN
JAKE "THE SNAKE" ROBERTS
"MACHO MAN" RANDY SAVAGE
ULTIMATE WARRIOR
ANDRE THE GIANT
DUSTY RHODES
RAZOR RAMON
YOKOZUNA

Hasbro's line of WWF figures took the world's most popular and powerful professional wrestlers and turned them into action figures, each with its own specialized wrestling move and a personalized bio card. The over-the top world of WWF wrestling, with its frequent televised events and outspoken and colorful kings of the ring, made it easy for Hasbro to develop a full line of figures.

HULK HOGAN/$10–$75

JAKE "THE SNAKE" ROBERTS/ $10–$75

"MACHO MAN" RANDY SAVAGE/$10–$75

ULTIMATE WARRIOR/$10–$75

Each wrestler is between 4½" and 5" tall, with powerful legs and thick, muscular torsos. The releases in the first two years came on a color-coded card with an "autographed" picture of the personality. The red-and-blue cards

were replaced in 1992 with different colored cards used for each new assortment of figures.

Hulk Hogan, the most popular figure, comes wearing a yellow-and-red tank top and briefs proclaiming

ANDRE THE GIANT/$95–$300

DUSTY RHODES/$95–$300

RAZOR RAMON/$10–$75

YOKOZUNA/$10–$75

"Macho Man" Randy Savage wears goggles and red briefs with white stars and features an "Elbow Smash" move. The figure is valuable, but not nearly as valuable as the large Andre the Giant figure that wears a one-shouldered singlet and wields a "Giant Jolt." The Dusty Rhodes figure wearing black with yellow polka dots is the most valuable of the first assortment of figures.

The 1992 Razor Ramon figure comes packaged on a yellow card, and the following year's Yokozuna comes on a red card. Yokozuna is larger than the other figures and features "Sumo Smash" action.

"Hulk Rules," and features "Gorilla Press Slam" as his action feature. Jake "The Snake" Roberts comes wearing teal-and-purple pants and a plastic python that wraps around his neck but doesn't take part in his "Python Punch."

Tragedy Sometimes Increases Value

WWF Wrestling has been a valuable franchise for LJN, Hasbro, and JAKKSPacific. Each time a company thinks the license has run out of steam another comes along and finds new life left in the category. Since **WWF Wrestling** figures have been produced for nearly 20 years, several of the ring stars have died. Thus far this has tended to add value to figures in their likeness.

The death of Owen Hart in his role of the Blue Blazer has caused this figure's value to skyrocket.

The only Hart Blue Blazer figure produced before his accidental death, seen on national pay-per view TV May 23, 1999, was in a Wal-Mart exclusive assortment. Owing to this unfortunate incident, the demand for the figure pushed the value to $90 within a month of first becoming available at retail.

WWF SUPERSTARS

JAKKS PACIFIC, 1996
SIZE: 6"
VALUE RANGE: $8–$12
(EXCEPT "DIESEL")

DIESEL
BRET "HIT MAN" HART
SHAWN MICHAELS
THE UNDERTAKER
STONE COLD STEVE AUSTIN

Two years after Hasbro stopped manufacturing figures, Jakks Pacific returned with an all-new line of WWF Superstars. The six original figures in the line featured joints

DIESEL/$15–$30

BRET "HIT MAN" HART

with "bone-crunching sound," which could be heard when the knee or elbow joints were bent forward or backward. Each figure

is packaged on cards with a quotation from each written across the bottom. The original six figures, including Diesel, Bret "Hit

SHAWN MICHAELS

THE UNDERTAKER

STONE COLD STEVE AUSTIN

Man" Hart, Shawn Michaels, and the Undertaker were so in demand that the stock of the first run sold out. New figures were rushed out on new, updated packaging without the quotations. The "Stone Cold" Steve Austin figure comes on the new card with a display base.

X-MEN

TOY BIZ, 1991–97
SIZE: 5"
VALUE RANGE: $5–$20

KILLSPREE
THE BLOB
MAGNETO
SABRETOOTH
GAMBIT
SPY WOLVERINE
CYCLOPS
PROFESSOR X
ROGUE
PHOENIX
BEAST
ELECTRONIC TALKING WOLVERINE
SENTINEL
WHITE QUEEN
JEAN GREY SPACE RIDERS
APOCALYPSE
BISHOP
STORM SAVAGE LAND
STORM ROBOT FIGHTERS

KILLSPREE

This line of action figures was first released by Toy Biz as The Uncanny X-Men, which was changed to X-Men when Marvel took over. The sculpting of the figures improved with the change in name, and the line grew rapidly.

THE BLOB

MAGNETO

SABRETOOTH

The X-Men are a group of Mutants, each born with a unique and extraordinary characteristic. The X-Men were brought together by Professor Charles Xavier, who trains them to use their superhuman powers for good—to fight for harmony and against prejudice. The archenemy of the X-Men is Magneto, who seeks to enslave mankind and create a new world order dominated by him and his followers, the Evil Mutants.

There is also an X-Men X-Force line, a spin-off from X-Men featuring characters with a little more of an edge. The powerful purple-and-

GAMBIT

SPY WOLVERINE

CYCLOPS

PROFESSOR X

ROGUE

PHOENIX

yellow figure of Kill-spree has "slashing blade arms." The Blob, also part of the X-Force line, is a very heavy figure featuring a "rubber blubber belly."

Magneto was manufactured as part of the original Uncanny X-Men line. Each figure had a special feature and came with a collector's card. Magneto, who uses the power of magnetism to achieve his evil ends, comes with magnetic hands and a

magnetic chest and is accessorized with "metal debris" that attaches to the figure's hidden magnets. Sabretooth combines human and tiger char-

BEAST

ELECTRONIC TALKING WOLVERINE

SENTINEL

WHITE QUEEN

JEAN GREY SPACE RIDERS

acteristics and has self-healing wounds. Gambit is dressed in a plastic cape, and the tall, lean figure has "power kick action."

Spy Wolverine was a 1993 Kay-Bee Toys exclusive version of the popular Wolverine character. He is dressed in black with gold body armor and is accessorized with many knives as part of his "thrusting knife action" feature. Also, in 1993 the figure of one-eyed Cyclops was

changed to a beefier figure with "light up optic blast" capabilities, and the X-Men leader, Professor X, appeared for the first time.

The female Rogue figure appears on the revised X-Men card with a "power upper-cut punch." Phoenix with "fiery phoenix power" is part of the

APOCALYPSE

BISHOP

STORM SAVAGE LAND

five-figure Phoenix Saga assortment, which came on two different cards.

Beast is one of the X-Men Classics figures,

STORM ROBOT FIGHTERS

a muscular blue figure packaged on a Classics card with "mutant flipping power." Electronic Talking Wolverine comes on a card advertising his talking feature.

A giant 14" Sentinel robot playset was manufactured to interact with the 5" X-Men figures, with special features like a chest plate that blows off to reveal a mutant prisoner, and a figure-gripping claw hook that retracts. Sentinel

also has knee targets that eject his boots when hit, causing the robot to topple over.

White Queen is part of the Generation-X assortment of figures that come packaged on specialized cards with accessories like White Queen's psychic energy spear.

The X-Men series continued to branch out into new assortments, like Missile Flyers, including revamped Apocalypse and

Bishop figures, and Space Riders, including a Jean Grey figure that powered the enclosed Hyper Jet craft. The popular Storm figure came in many forms throughout the line, including a white-clad Storm in a Savage Land window-box set, a mutant dinosaur called Colossus, and a black-clad Robot Fighter Storm with a spinning weather station.

ZORRO

GABRIEL, 1982
SIZE: 3¾"
VALUE RANGE: $5–$20

ZORRO

The figure of Zorro, legendary champion of justice, is part of a six-figure line based on a Saturday morning animated TV show about the swashbuckling hero. Each figure included a plastic sword and a pistol, and they were packaged on orange-and-yellow cards with a picture of Zorro and his horse. Zorro is dressed all in black, with a red sash and a real red cloth cape.

ZORRO

INDEX OF FIGURES

B

TYRANNOSAURUS REX with ELECTRONIC Roar & Stomping Sound!

K

L

M

N

O

S

SGT.

TRANSFORMS FROM HELICOPTER
TO ROBOT AND BACK!

T

X

Y

Z